Green Chemistry and Sustainable Technology

Aims and Scope

The series *Green Chemistry and Sustainable Technology* aims to present cutting-edge research and important advances in green chemistry, green chemical engineering and sustainable industrial technology. The scope of coverage includes (but is not limited to):

– Environmentally benign chemical synthesis and processes (green catalysis, green solvents and reagents, atom-economy synthetic methods etc.)
– Green chemicals and energy produced from renewable resources (biomass, carbon dioxide etc.)
– Novel materials and technologies for energy production and storage (bio-fuels and bioenergies, hydrogen, fuel cells, solar cells, lithium-ion batteries etc.)
– Green chemical engineering processes (process integration, materials diversity, energy saving, waste minimization, efficient separation processes etc.)
– Green technologies for environmental sustainability (carbon dioxide capture, waste and harmful chemicals treatment, pollution prevention, environmental redemption etc.)

The series *Green Chemistry and Sustainable Technology* is intended to provide an accessible reference resource for postgraduate students, academic researchers and industrial professionals who are interested in green chemistry and technologies for sustainable development.

More information about this series at http://www.springer.com/series/11661

Mario Malinconico
Editor

Soil Degradable Bioplastics for a Sustainable Modern Agriculture

 Springer

Editor
Mario Malinconico
Institute for Polymers, Composites
    and Biomaterials (IPCB-CNR)
Pozzuoli
Italy

ISSN 2196-6982   ISSN 2196-6990   (electronic)
Green Chemistry and Sustainable Technology
ISBN 978-3-662-57181-1   ISBN 978-3-662-54130-2   (eBook)
DOI 10.1007/978-3-662-54130-2

Printed on acid-free paper

This Springer imprint is published by Springer Nature
The registered company is Springer-Verlag GmbH Germany
The registered company address is: Heidelberger Platz 3, 14197 Berlin, Germany

Preface

When in 1948 Prof. E.M. Emmert built, for the first time in the world, a greenhouse covered with plastic sheets (cellulose acetate film), maybe he did not imagine that a real revolution in the agriculture world was starting. Professor Emmert used, for his first trial, four ft square plastic films that replaced the glass sheets, employed for traditional greenhouses, up to that time used just in top botanical gardens, to grow and study tropical plants and flowers. Initially, his goal was to realize a new greenhouse made with innovative and cheaper materials. The experimental results were so appealing to extend the use of plastic film for mulching and low tunnel. Later, Prof. Emmert moved to a more efficient polyethylene film. Thanks to his studies and his great contribution to agriculture, he is worldwide considered "*the father of plastic greenhouse.*"

Ever since, the so-called plasticulture has extended and brought many important benefits to modern agriculture, among which it is necessary to highlight the reduction of water consumption and loss of minerals, the reduction of use of chemicals for spontaneous weed control, the possibility to manipulate light, to thermally insulate the crops, to provide mechanical protection.

Where are we today? About 5 million tons of agricultural plastic resin is used worldwide and the number is growing. Films' application has been extended to other items, such as sheets, rods, tubing, and transplanting pots. Burning these plastics in the fields is not an option because it contributes to serious environmental air and particulate pollution. Therefore, this practice is being phased out and many countries have strict regulatory bans on plastic film burning. Collection, cleaning, and recycling these plastics to same or other products offers an approach to managing plastic waste and there are several companies that offer these services. Nevertheless, a complete analysis of the managing and running costs of collections, grinding, cleaning, and recycling plastics shows that, at least for some items, material recycling is economically and environmentally unsustainable, and soil degradation would be a feasible and desirable managing option.

Although agriculture soil degradable plastics have still a less than one digit share of the market of plastics, they are growing at very fast rate, and the properties of biodegradable compostable plastics have even opened new fields of applications which were not possible with polyolefin-based plastics (for example, soil degradable nursing, and transplanting pots).

This book originates from at least 30 years' experience of soil degradable plastics for agriculture. The seven chapters span from films for mulching, direct cover, and tunnel to other applications. Some chapters open windows to future technologies, such as biodegradable waterborne varnishes, which are still far from technological maturity. The authors, to whom goes my gratitude, are among those who made most of their research efforts, in academia and in public and private research centers, to design, process, test, and optimize the plastics, either biobased or synthetic, and to assist the development of norms and directives. Harmonized environmental norms and directives are absolutely necessary. We must be conscious that a man-made material which is designed to remain in the nature after use poses ethical problems on the top of the reasons of technology, economy, and profit.

This book is dedicated to Rosario Palumbo and Gianni Maglio, retired Professors of Chemistry at University of Naples Federico II, and to Alfonso Maria Liquori, Professor of Chemistry at University of Rome Tor Vergata (deceased). By their teaching and their moral rectitude, they have strongly contributed to my human and scientific personality.

Naturam expelles furca tamen usque recurret (**Orazio,** *Epist.*, **I, 10, 24**)

Pozzuoli, Italy   Mario Malinconico

Contents

Contributors

Giorgio Borreani DISAFA, University of Turin, Turin, Italy

Demetres Briassoulis Agricultural Engineering Department, Agricultural University of Athens, Athens, Greece

Alessandro K. Cerutti Department of Agriculture, Forestry Food Science, University of Turin, Turin, Italy

Joan Costa Department of Horticulture, Botany and Gardening, ETSEA University of Lleida, Lleida, Spain

Francesco Degli Innocenti Novamont S.p.A., Novara, Italy

Sara Guerrini Novamont, Novara, Italy

Barbara Immirzi Institute for Polymers, Composites and Biomaterials (IPCB-CNR), Pozzuoli, NA, Italy

Mario Malinconico Institute for Polymers, Composites and Biomaterials (IPCB-CNR), Pozzuoli, NA, Italy

Lluís Martín-Closas Department of Horticulture, Botany and Gardening, ETSEA University of Lleida, Lleida, Spain

Pasquale Mormile Institute of Applied Sciences and Intelligent Systems "E. Caianiello" (ISASI) of CNR, Pozzuoli, NA, Italy

Ramani Narayan Michigan State University, East Lansing, MI, USA

Ana M. Pelacho Department of Horticulture, Botany and Gardening, ETSEA University of Lleida, Lleida, Spain

Francesco Razza Novamont, Terni, Italy

Gabriella Santagata Institute for Polymers, Composites and Biomaterials (IPCB-CNR), Pozzuoli, NA, Italy

Giacomo Scarascia Mugnozza Department of Agricultural and Environmental Science (DISAAT), University of Bari, Bari, Italy

Evelia Schettini Department of Agricultural and Environmental Science (DISAAT), University of Bari, Bari, Italy

Noam Stahl Ginegar Plastic Products Ltd., Kibbutz Ginegar, Israel

Henk Voojis Amsterdam, The Netherlands

Giuliano Vox Department of Agricultural and Environmental Science (DISAAT), University of Bari, Bari, Italy

Chapter 1
The World of Plasticulture

Pasquale Mormile, Noam Stahl and Mario Malinconico

Abstract Since its appearance in agriculture, plastic films have revolutionized this sector with huge benefits in terms of quality and quantity of crops. In the last decades, agricultural films gained not only a great deal of interest and attention but also a big market more and more extended in any country of Europe, America, and Asia, with a constant positive trend that does not know crisis. Plasticulture is a term that indicates the world of plastic films applied in agriculture, ranging from greenhouse covers to mulches, from low tunnels to solarization films, from Totally Impermeable Films (TIF) to photo-selective films. It includes raw material (several kinds of polymers), different types of films, applications, agronomical performances, and the recycling problems. In this chapter, a review of Plasticulture is presented, from historical origins to the films extrusion technique, from their physical properties to the different typologies and uses. In particular, a new generation of agricultural films is described with characteristics, performances, and employment modalities. Finally, the new frontier of plastic films for greenhouse, with open window to UV-B radiation, is presented, considering the huge potentiality for improving further the crops quality and the increase of different nutraceutical contents in fruits and vegetables.

Keywords Agricultural films · Greenhouse covers · VIF · Low tunnel · Mulches · Photo-selective films · Solarization · UV-B radiation

P. Mormile (✉)
Institute of Applied Sciences and Intelligent Systems "E. Caianiello" (ISASI) of CNR, Pozzuoli (NA), Italy
e-mail: p.mormile@isasi.na.cnr.it

N. Stahl
Ginegar Plastic Products Ltd, 36580 Kibbutz Ginegar, Israel
URL: http://www.ginegar.com

M. Malinconico
Institute for Polymers, Composites and Biomaterials (IPCB), 80078, Pozzuoli (NA), Italy
e-mail: mario.malinconico@ipcb.cnr.it

© Springer-Verlag GmbH Germany 2017
M. Malinconico (ed.), *Soil Degradable Bioplastics for a Sustainable Modern Agriculture*, Green Chemistry and Sustainable Technology,
DOI 10.1007/978-3-662-54130-2_1

1

1.1   Introduction

When in 1948 Professor E.M. Emmert built, for the first time in the world, a greenhouse covered with plastic sheets (cellulose acetate film) [1], maybe he did not imagine that a real revolution in the agriculture world was starting. In that time, based cellulose acetate materials were spreading across different industrial sectors, including cinema, medicine, automotive, military, clothing, and furnishing. Due to the chemical and physical characteristics, plastic products replaced quickly those made with traditional materials, like metal or metallic alloys, natural fibers, wood, glass and paper. The main futures of plastic materials were colourability; sound, thermal, electrical, mechanic insulation; resistance to chemical corrosion; water, molds and bacteria repellency; easy processing and cheaper costs. No other material offered the same advantages.

Professor Emmert used, for his first trial, four ft square plastic films that replaced the glass sheets, employed for traditional greenhouses, up to that time used just in top botanical gardens, to grow and study tropical plants and flowers. Initially, his goal was to realize a new greenhouse made with innovative and cheaper materials. The experimental results were so appealing to extend the use of plastic film for mulching and low tunnel. Later, Prof. Emmert moved to a more efficient polyethylene film. Thanks to his studies and his great contribution to agriculture, he is worldwide considered "*the father of plastic greenhouse*".

With the advent of polyethylene, Plasticulture, a term refers to the use of plastic materials in agriculture (mainly mulch, low tunnel, greenhouses cover, solarisation film, fumigation film, and packaging), burst [2]. Plasticulture spread across the world in a short time, from USA to Europe and Far East [3–5]. The Plastic films consumption increased exponentially year-by-year [6, 7], reaching the current global shares of 3.9 million tons, mainly in Asia (roughly 70%) and Europe (16%).

The broadest expansion of greenhouses in the world is located in the Far East (China, Japan, and Korea) with a share of 8%; in the Mediterranean area the percentage reaches 15% of global share. The trend is constantly positive. Recently, the use of agricultural plastic films in Middle East and Africa has been increased by 15–20% per year. Today, the biggest consumption of plastic film is in China, with an annual growth of 30%, and a total volume of 1,000,000 t/year. A recent worldwide estimation confirms that agricultural grounds are covered in this way: 18,000,000 ha by mulching, 920,000 ha by low tunnels and 1,300,000 ha by greenhouses.

Plastic materials offer a wide range of applications, from mulching to low tunnels, from solarisation films to greenhouses covering, from barrier films to silage films, from irrigation systems to "functional" films, from packaging to flowerpots [8, 9]. Figure 1.1 shows the consumption of agricultural plastic films in the world, forecasted for 2015, while Fig. 1.2 reports schematically the global distribution of agricultural plastic films consumption forecasted for the same year [10].

The basic material, for most of the products, is the polyethylene (PE) in the low density version (LDPE) and linear low density version (LLDPE). The chemical

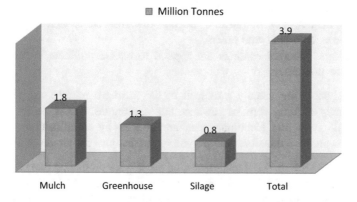

Fig. 1.1 Consumption of agricultural films in the world, forecasted for 2015

Fig. 1.2 Geographic
distribution of agricultural
plastic films consumption,
forecasted for 2015

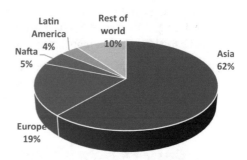

formulations adopted for the products mentioned above include different elements
(additives and stabilizers), used on the basis of the optical, mechanical and thermal
proprieties required for the different kinds of films, according to the specific use.

1.2   A Brief Introduction to Production Techniques: From Raw Material to Co-extrusion

The raw materials that are used to produce agricultural plastic tools may be divided
into three groups:

(1) Polymers. These are the main building blocks of the product, i.e., the "Plastic"
    (or more accurately—the Thermoplastic Polymers) materials. The most
    common polymer used in plasticulture is Poly Ethylene (PE) and its deriva-
    tives. Other common polymers used for plasticulture are Poly Vinyl Chloride
    (PVC); Poly Propylene (PP); Poly Carbonate (PC).

(2) Additives. Chemicals that are blended into the polymer and modify the properties of the product are called "Additives". In plasticulture they include anti-oxidants, UV stabilizers, IR absorbers/reflectors, anti-drip, anti-mist, biocides, colorants, and others.

(3) Coatings. External coatings are applied to plastic products to modify their surface properties.

The polymers are usually produced in the form of granules or pellets. The additives may be in the form of powders, flakes, granules, waxes, or liquids. There are several production techniques that are utilized by manufacturers of plastic products in order to transform the polymer granules into the shape of a product:

(1) Injection Molding. It is a manufacturing process for producing parts by injecting material into a mold. Material for the part is fed into a heated barrel, mixed, and forced into a mold cavity, where it cools and hardens to the configuration of the cavity. This production technique is used for the production of singular (non-continuous) parts such as clips, drippers, filters, flowerpot, and so on.

(2) Extrusion. In the extrusion process, molten plastic is pushed through a die, which shapes it. Then the plastic is cooled, and either cut into sections or rolled up. Extrusion is a continuous process capable of making parts of any length.

Some examples of extruded parts include films (greenhouse covers, mulch), tubing, yarns, and sheets. Modern extrusion lines are utilizing a technology known as "Co-Extrusion". In this process the shaping die is fed by more than one extruder simultaneously (Fig. 1.3).

Fig. 1.3 Co-extrusion of a five-layer film for greenhouse cover the result is a multi-layer product, i.e., a product that is structured of several discrete layers, each made of a different mixture of materials and hence each layer has different properties. For example, modern greenhouse covers are typically 3–5 layer films. The outer-most layer is designed to have the best UV resistance and to reduce dust attraction. The inner-most layer is designed to prevent formation of water droplets (anti-drip) and the middle layers contribute most of the mechanical strength of the film

1.3   The Active Role of a Plastic Film: Micro-climate Management

The primary role of plastic films as an agro-technical tool is to allow the grower to better control the micro-climate underneath the film.

This role is achieved by exploiting some inherent properties of the plastic films, as well as specially designed properties. In this paragraph we will highlight the main functions of an agricultural film. In the next paragraphs these functions will be explained in more detail.

- *Mechanical protection.* The continuous film forms a barrier that prevents mechanical damages to crops. Factors such as animals, hale, rain, dust and others are kept outside the covered area.
- *Thermal insulation.* Thermoplastic polymers have low thermal conductivity, which makes them good insulators. In addition, the film is reducing air circulation inside the covered space. As a result, the temperature under the film is higher than the ambient temperature. Thermal additives (such as Mid-IR absorbers) are also used to reduce heat losses.
- *Manipulation of light.* As a general rule, in most cases the crops under an agricultural film require high levels of Photosynthetically Active Radiation (PAR) [11]. This is translated into a requirement from cover films to allow as much PAR into the structure as possible. In other cases, the films may be used to manipulate the quality of light under (or above) the films.

1.4   Technical Characteristics: Mechanical, Optical and Thermal Properties

A plastic film used for agriculture is characterized by three main properties: mechanical, optical, and the thermal one. The producer and the end-user define the desired properties according to the function that the film has to fulfill [12–14].

- *Mechanical properties*:

  (1) *Tensile strength*—measures the force needed to pull the material until it yields or breaks (per unit area). Products that are under mechanical loads such as weight, pull, compression, or wind require high tensile strength.

  (2) *Elongation*—measures the extent to which a material stretches until it yields or breaks when pulled. Usually it is reported as (%) percentage of original dimensions. Also it serves as an indicator of the flexibility of a product. Brittle materials show low elongation at break point. Ductile materials have higher elongation. Products that are expected to change their form to allow for better fit to the terrain where they are used should have high elongation rates. A good example is mulch films.

(3) *Impact strength*—is the amount of energy required to fracture a material; it is a measure of the material's resistance to mechanical shock. Products that are subjected to sudden loads, such as wind gusts, should have high impact strength.

(4) *Flexural strength*—of a material is defined as its ability to resist deformation under load. In other words, how much force is needed to bend a product before it yields or breaks. Fittings and profiles should have high-flexural strength.

(5) Tear propagation—the average force required to propagate a single-rip tear starting from a cut in a film.

- **Optical Properties**:

(1) *Color*—most plastic products used in agriculture are either colorless or black. Some unique products may have chromatic colors. The color is used for marking (i.e., colored clips used to mark flowers) or for technical purposes (such as colored mulches).

(2) *Total light transmission*—measures the percentage of light that will pass through a film or a net. It is common to report the total amount of transmitted light in the UV-Vis-NIR range (280–2500 nm). For greenhouse covers the value should be greater than 85% and is typically 88–92%. In some cases shading is required and the light transmission is reduced to much lower values.

(3) *Diffused light transmission*—measures the percentage of light that passes through a film and that is scattered from the incidence plane by 5° or more. Greenhouse covers may have low (10–20%); Medium (20–40%) or high (above 45%) diffused light transmission.

(4) *Spectrum of transmitted light*—Some films and nets are designed to alter the spectrum of light that reaches the crop and its surrounding. Special additives are employed in order to absorb and/or reflect selectively specific bands of wavelengths. The result may be, for instance, the reduction of the amount of UV light in the greenhouse while keeping high levels of PAR. Colored nets represent another example that manipulates the quality of the solar radiation under them.

(5) *Spectrum of reflected light*—the spectrum of light that is reflected from a film or a net impact the crops above them. It may also influence the behavior of insects approaching the crops. Some colored mulches are known to interfere with the activity of insects such as thrips and white flies.

- **Thermal Properties**:

(1) *HDT* (heat flection temperature)/Ultimate service temperature—a measure of a product's ability to remain useful at high service temperature [15].

(2) *Thermicity*—is the parameter describing the capacity of an agricultural film to accumulate heat under a greenhouse. Its value (%) is obtained from the measurement of the transmission of IR radiation through a film, by calculating the integral of the spectral curve in the interval between 7 and 14 μm.

The lower the thermicity is, the lower the transmission of IR and the lower the heat loss during the night. The use of copolymers (such as EVA and EBA) and the use of IR additives (such as talk, caolin, china-clay, and others) allow film producers to control the thermicity of greenhouse covers.

1.5   Functional Aspects: Anti-drip, Anti-fog, and Anti-mist System

Greenhouse covers and tunnel covers may lose some of their initial properties due to condensation of water on the inside surface of the film. The air inside the greenhouse has high relative humidity due to irrigation and respiration. The film surface is usually cooler than the air inside, leading to condensation. Since the films are usually made of PE with low surface energy (30–32 dyne/cm), the water tends to form hemispherical droplets. The shape of water droplets on the internal surface of greenhouse covers is referred to as "Fogging" or "Dripping". It has many negative aspects like reduction of light transmission and formation of excessive moisture on leafs and flowers.

In order to prevent this form of condensation, film producers are using Anti-drip additives (sometimes called Anti-fog). These are surface-active chemicals that increase the surface energy of the film. As a result, the condensation water tends to form a continuous layer on the film, thus avoiding the negative effects of discrete droplets.

1.6   Agricultural Films for Mulching, Low Tunnel, and Greenhouse

The range of plastic films used in agriculture is wide; it varies according to the cultivation requirements, the agronomic practices, the specific use, the local clime, and the geographical conditions. Below the most common types of agricultural films are described.

Mulch. The practice of mulching as soil coverage with various materials to prevent the growth of weeds is very old. Since the ancient times (2500–3000 years Before Christ), people from several countries (Egyptians, Etruscans, Persians, Romans) adopted the mulching technique. They just simulated the leaves and woody debris sediments on the soil, which prevent the weed growth because of the combined effect of the solar radiation block and the biochemical action of the decomposition of organic materials. *We did not invent mulch. Mother Nature did.*

Before the use of plastic films, many different materials were employed for mulching: straw, dried leaves, bark of trees, cardboard, gravel, lapilli, jute, cocoa husks, natural fibers (mainly coconut and hemp) and organic waste.

Plastic mulching films gained briefly wide attention from the agricultural world. They keep the soil wet, block the solar radiation to hinder the weeds growth and protect the soil from the erosion of the pouring rain. Plastic films are also easy to put on the soil with some special machines, covering tens hectares in a short time. The films are supplied according to the different needs of the customers (several lines of holes with different sizes).

Until a few years ago, the average thickness of a mulching film was 50–70 μm; today, thanks to the techniques of co-extrusion, the thicknesses are largely reduced (20–30 μm), while still providing good mechanical properties. For years, farmers used almost exclusively the black film, which is able to totally block the solar radiation, or the transparent one, which is able to heat the soil, but often renouncing a satisfactory mulching effect [16]. Recently, coloured films (white, green or yellow) and photo-selective film are penetrating into the mulching practice with a great success and satisfaction, thanks to the agronomical performance that they guaranty besides the mulching effect. In Fig. 1.4 different kinds of mulching films are shown. This new type of mulching films will be treated in the next paragraphs.

Low tunnel. Transparent films, clear or diffused light, have a width ranging from 1.2 to 6 m, and a thickness from 30 to 150 μm. The small size version (film width from 1.2 to 3 m) of this type of film is widely used for growing both in open field and in the greenhouse, for watermelons, asparagus, and vegetable crops. The larger version (3–6 m) is widely used for strawberries, melon, and vegetable crops. Depending on the needs of the growing areas, the type of product can be supplied with high thermal characteristic. Figure 1.5 shows the common use of low tunnels in open field for watermelon cultivation. In this photo it is possible to compare also the difference between low tunnels with and without anti-dript system.

Fig. 1.4 Photo-selective mulch films at work. A recent trial aimed to test different kind of mulch films

Fig. 1.5 Low tunnel in open field used for watermelon cultivation

Greenhouse cover. This is the most widely used plastic film for protected cultivation. There is a wide range of products, especially in recent years, which are
able to satisfy almost all needs of farmers. Film sizes depend on the type and
dimension of structure and they have widths ranging from 6 to 14 m, while the
thicknesses (ranging from 100 up to 200 µm) are proposed according to the
duration of the films. The production technique is based on the extrusion or
co-extrusion that allows to obtain multi-layers film up to five layers, ensuring high
mechanical properties and the use of ad hoc formulations for functionalised films.
The typology is very varied: direct light version (clear film) or diffused light version
(opaque film), it ranges from simple version with just stabilizers to thermic films,
from highly specialized films to functional covers as photo-selective films. Through
special additives, and thanks to the co-extrusion technique, it is possible to fabricate
film with desired optical, thermal, and mechanical properties in order to optimize
the agronomical performance for all types of growth [17, 18]. According to crop
needs, a cover film is supplied with different functions like anti-drip, anti-dust and
anti-fog system. Furthermore, it is possible to employ:

– cover films which cut UV radiations or that are partially transparent to UV;
– coloured films (to induce photomorphogenesis in plants) or with high reflectivity to IR component of solar radiation (cooling effect) and, finally,
– functionalized films for supporting farmers aimed to specific agronomical performances (UV-B induced secondary plant metabolites) [19].

Figure 1.6 shows high-density greenhouses, which is common in many sites
around the world.

Fig. 1.6 High density greenhouse area

1.7 Solarisation Film

Soil solarisation is a well-known agronomical practice, based on the use of special plastic films, which are employed to sterilize the soil by heating it through solar energy [20]. A good solarisation is able to increase the soil temperatures at different depths up to eliminate the most part of pathogens accumulated during any crop cycle. This natural soil sterilization depends strongly on the temperature levels obtained, at different soil layers, during the solarisation process. In other words, the higher the temperature the more efficient the solarisation effect is [21, 22].

The traditional solarisation practice requires the use of a plastic film with very special optical and thermal properties, covering a soil which has been well wet for a long enough period of time (4–5 weeks) in order to obtain satisfying soil steril-ization. Nevertheless, the use of inadequate plastic film could frustrate good soil sterilization, because the temperature average level during the whole period, esti-mated to be about 42 °C, is under the minimum threshold to eliminate very harmful pathogens, like Nematodes.

In Fig. 1.7a, b it is shown a comparison between two solarisation plastic films with different thermal and optical characteristics that affect the final results.

In such figures, the temperature average level (42 °C) is plotted, under which the soil sterilization effect is insufficient. Figure 1.7a shows an unsatisfying situation, due to a shoddy plastic film, where most of the temperatures at different depths are under threshold, while Fig. 1.7b exhibits a better situation, thanks to more appro-priate optical and thermal properties of the employed film for solarizing.

All this depends heavily on the quality of the plastic film used and on the right method adopted.

(a)   **(b)**

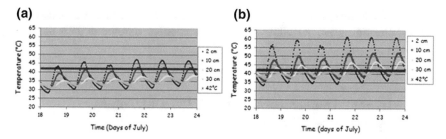

Fig. 1.7 Comparison of experimental results obtained with two different plastic films. In **a** the data exhibit an unsatisfying solarisation, while in **b** the data present temperatures more suitable for soil sterilization

A good film for solarisation has optical characteristics aimed to enhance the "greenhouse effect" in the soil, trapping in it the heat associated with solar radiation. This means that the film should be as transparent as possible to solar radiation (200 nm up to 2500 nm), and opaque to IR radiation emitted from the soil. Under these conditions the heat is accumulated in the soil and temperatures at depths up to 35–40 cm increase until a satisfactory sterilizing effect in soil is achieved: elimination of "negative" pathogens of animal origin (e.g., Nematodes) or vegetable origin (various kinds of fungi and molds).

Therefore, as shown in Fig. 1.8a, b, the ideal optical properties of a good solarisation film exhibit very high transitivity in the range of solar radiation and a considerable block of IR radiation (heat energy) emitted from soil. This property, in theory, induces a heat accumulation in the soil that turns in an increase of temperature at different depths.

Recently, solarization technique has been implemented by simulating the thermal solar panel effect, which is used for hot water supplying. Thanks to this hybrid innovative method, it is possible to increase further the temperature at different depths in the soil, up to 8 °C [23, 24]. The solarisation films have widths ranging from 6 to 12 m, and a thickness between 25 and 40 μm. They are produced with the

(a)   **(b)**

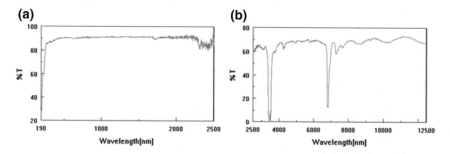

Fig. 1.8 Optical spectra of a film well performing in UV-Vis-NIR region (**a**) and IR region (**b**). The total transparency at solar radiation associated with two absorption peaks in IR region indicates ideal characteristics for solarisation

co-extrusion technique and the most part of films presented into the market possesses anti-drip system to allow the greatest possible transparency to the solar radiation.

1.8 Virtually Impermeable Film (VIF) and Totally Impermeable Film (TIF)

The barrier film is known under the trade name VIF (Virtually Impermeable Film) because it serves to block the passage of harmful gases (barrier effect) during the practice of fumigation. This practice consists of the injection in the soil of very aggressive chemicals for the removal of pathogens present in the soil in the post-harvest phase. The films cover the soil surface to minimize the transfer of harmful gases, which are often toxic to the environment, in order to prevent damage to farmers and the environment [25, 26]. The most important feature that distinguishes a barrier film is the so-called "permeability" which is measured in g/m^2h. A good film has a permeability barrier that does not exceed 0.02 g/m^2h. The lower the permeability of a VIF is, the better the barrier effect is. Recently, a new generation of barrier film called TIF (Totally Impermeable Film) has been proposed with more advanced features with regards to the permeability, lowered up to 20 times of its average, and with a better heat capacity. The most critical aspect of a barrier films is linked to temperature increase under operating condition, which determines, as a result of thermal expansion, a drastic increase of permeability. With the advent of TIF this problem is greatly reduced. It is clear that it is not necessary to provide transparent VIF, or TIF, because their function is only to contain gas during and after fumigation. In fact, market offers VIF also in black or coloured (Brown and silver) versions, which remain after fumigation on the soil to be used as mulching film.

1.9 The Role of a Plastic Film in Photosynthetic Process

Solar radiation is the radiation that comes from the Sun in the form of electromagnetic waves that cover the spectrum ranging from 0.2 to 2.5 μm. In particular, the solar radiation is constituted of ultraviolet rays (UV), visible region (Vis) and an infrared (IR) part, here it is schematically reported:

- 0.2–0.38 μm: ultraviolet-UV (6.8% of total energy)

  - 0.2–0.28 μm > UV-C
  - 0.28–0.35 μm > UV-B
  - 0.35–0.38 μm or > UV-A

- 0.38–0.78 μm: visible-Vis (48%)
- 0.78–2.5 μm: infrared-IR (45.6%)

The contribution of solar radiation as an energy source to the plant system is crucial. Without this energy any living form on our planet would be impossible. The three essential parts of the solar radiation (UV, Vis and IR) have a different role but are substantial for the plant world:

(a) UV rays affect the coloring of plants including fruits (apples, nectarines, cherries, etc.) and flowers, and they determine the fungal sporulation. Furthermore, UV rays induce the secondary metabolites in the plants. Most of the insects "see" and orient thanks to UV rays;
(b) the visible part of the solar radiation is essential for photosynthesis and is the main component along with water and CO_2. Photosynthesis gives rise to the sugar chain (primary metabolites), which is the fundamental for the plants growth;
(c) IR rays (infrared) represent the hot component of radiation, which is essential for the vegetative state.

The portion of the electromagnetic spectrum used to promote the processes of plant development, is between 400 and 700 nm; this range is called Photosynthetically Active Radiation (PAR) and represents the portion of solar radiation giving rise to photosynthesis.

Through this process, the plant takes CO_2 (carbon dioxide) from the air and uses the required carbon, in turn releases into the environment water and oxygen. This bio-chemical process is powered by absorption by plants of electromagnetic radiation PAR. Proper development of a plant depends on the right balance between the two processes of *photosynthesis* and *respiration* (chlorophyll synthesis). Photosynthesis involves the absorption of carbon dioxide and release of oxygen. A balanced development occurs only when the action of photosynthesis prevails over the respiration.

Photosynthesis is stimulated by blue radiation, at the wavelength of 425–450 nm, and red radiation, at the wavelength of 575–675 nm. The activation of the process of photosynthesis requires a minimum amount of light energy, which depends on the cultivar: some plants grow only if they are exposed to large amounts of light, (heliophilous plants such as the sunflower), while others only develop in the presence of low irradiation (ombrophyte plants such as avocado or yucca). In heliophilous plants, photosynthetic action grows with increasing light irradiation, but because the assimilation by plants is subject to saturation, it reaches its maximum value close to 10,000 lx.

In this context, a plastic film, seen as interface between solar radiation and the vegetable system (to cover greenhouses, tunnel or mulch), can play a decisive role if it can manage the solar radiation, which makes it an active element in the light-plant interaction: "photo-selective effect".

Photo-selective films are able to select the electromagnetic radiation both entering and exiting greenhouse, determining optical and climatic conditions more suitable to the growth [27]. This mechanism of solar radiation management is similar for *low tunnel, mulching* and *solarisation*.

Photo-selective mulch. The traditional black films used for mulching guarantee the mulching effect, thanks to the blocking of solar radiation because the filtering of PAR radiation prevent weed growth. Contrarily to traditional films, photo-selective mulching blocks only the PAR of solar radiation, by modulating the part useful to heat or cool the soil. These films have optical properties such as to select both incoming electromagnetic radiation (solar radiation) and outgoing electromagnetic radiation (radiation emitted from the ground) [22]. In particular, thermal films, blocking the PAR, permit the transmission of the "warm" part of the solar radiation (near Infrared and part of the Middle IR commonly known as SW-short wave) and reduce the passage of thermal radiation emitted from the soil (medium/far IR commonly known as LW-long wave).

The films with these features are present in the market with various colors such as brown, red or green and are used for winter transplants up to May depending on the thermal state of the soil and on the climatic conditions. The optical spectra of a "heating" film are shown in Fig. 1.9a, b, where the transmissivity is presented as a function of wavelength in the range of solar radiation (Fig. 1.9a) and in the range of radiation emitted from the soil (Fig. 1.9b).

In Fig. 1.9a, the transmittivity in the PAR is <5%, and this guarantees the "mulch effect", while it increases up to 80% in IR region (2500 nm). In Fig. 1.9b, two absorption peaks are evident, giving rise to the trapping of heat in the soil. As a consequence, the temperature increases at the roots.

Another type of photo-selective film works to cool the soil thanks to their high reflectivity to solar radiation, in addition to mulch. This property allows to avoid thermal damage during the transplants warm period. The optical characteristics of these films show a block at PAR and a high reflectivity in the NIR. As an example of spectrum of photo-reflective film, Fig. 1.10a, b shows the transmissivity as a function of wavelength of one of these films, which are present in the market with different color such as white, silver or yellow.

To complete the panorama of photo-selective mulching film, it is noted also the presence of simple transparent films, which are used to heat the soil without mulching effect since the PAR pass completely.

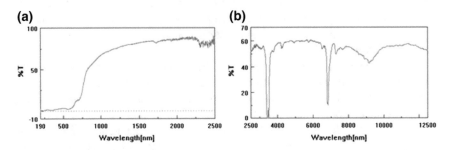

Fig. 1.9 Optical spectra of a thermal mulch film used for winter transplants in UV-Vis-NIR region (**a**) and IR region (**b**)

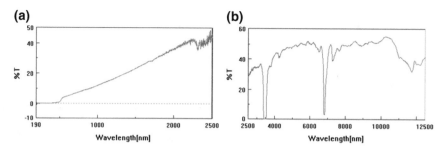

Fig. 1.10 Optical spectra of a photo-reflective mulch film used for spring and summer transplantsin UV-Vis-NIR region (**a**) and IR region (**b**)

Photo-selective low tunnel. Also for the films used as low tunnel, exists a range featuring photo-selective characteristics, which is able to guarantee the greenhouses effect. Thanks to the optical and thermal properties of these films, it is possible to obtain high thermal action in order to protect crops from a sudden drop in temperature that slows the growth of plants.

Greenhouse cover. In recent years, the photo-selective film for greenhouses cover penetrated into the market and are constantly innovated and improved, thanks to the active research in this sector. The physical principle that characterizes these films is the light management: selection of incoming electromagnetic radiation into the greenhouse (solar radiation) and the outgoing one (radiation emitted from the greenhouse ambient) [28].

Thanks to their optical and thermal properties, these films are able to offer a valid technical support to farmers according to their requirements in the cultivation. The types of photo-selective film for covering greenhouses ranges from films that enhance highly the "greenhouse effect" to those reflect partly the NIR radiation (cooling effect).

Other films completely block UV rays of sunlight, creating the dark in greenhouses for insects (the insects see with UV rays) and unfavorable conditions for sporulation of fungi. This mechanism creates the conditions for a drastic decrease of insects, that loss orientation and move outside the greenhouse. On the other side, it obtains a reduction of fungal diseases such as downy mildew, powdery mildew and botrytis. There are films that have an open "window" to the UV rays. In this case the film transmits UV light that improves the colour of fruits such as nectarines, peaches, apricots and cherries. More recently the interest is shifting to a further type of photo-selective cover, that transmits only a portion of UV rays, particularly UV-B. Recent studies have highlighted the regulatory properties of low, ecologically relevant UV-B levels that trigger distinct changes in the plant's secondary metabolism resulting in an accumulation of mainly phenolic compounds, carotenoids and glucosinolates. These interesting studies demonstrated the possibility to exploit UV-B induced metabolic changes in fruit, vegetable and herbs to satisfy consumer demand for natural health-promoting food products. The contribution of UV-B to inducing secondary plant metabolites, as potential benefits for human

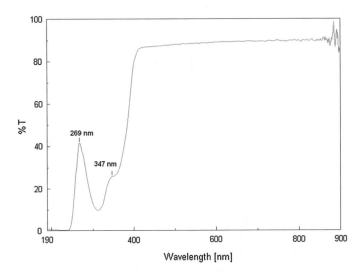

Fig. 1.11 Transmittivity as a function of wavelength of a greenhouse film, having a "windows" in UV-B radiation

health, is emphasized by new plastic films. It is because their optical properties (see Fig. 1.11) permit UV-B crossing into the greenhouse [19].

Other types of films for covering photo-selective greenhouses are represented by coloured film (red or blue) to induce the photomorphogenesis in the plants. These films, according to their color, behave as an optical filter that selects the visible light under greenhouse, affecting the morphology of the plant. A preponderance of light blue-violet may cause delays in height growth (growth-regulating effect), while an irradiation mainly based on infrared and red stimulates a longitudinal overgrowth.

1.10 Photo-Selective Films for Saving Water in Agriculture

Photo-reflective mulch films represent, in the panorama of agricultural films, a valid support for Spring and Summer cultivations, both in open field and under greenhouse. In fact, thanks to the high reflectivity of these films, thermal aggression, that causes serious problems to plants when traditional black mulch films are used, is avoided. Yellow or silver colored photo-reflective films protect plants from damages, assure the mulching effect, give a valid support to Integrated Pest Management and, according to recent trials, greatly contribute to saving water. This further advantage is determined by the high water condensation under the mulch film and this gives rise to reduction of irrigation. Water saving also means energy saving for electric system of water circulation. Trials performed at different

Fig. 1.12 Photo-reflective mulch films used for saving water

geographic and ambient context confirmed that the use of photo-reflective mulch films during the hot season allows to save water up to 30%. Recently, an experimental activity was performed aiming to demonstrate that photo-reflective films (yellow, silver, and withe) give the possibility to save water in agriculture [29]. The trial was performed in a greenhouse (Fig. 1.12) and several films were tested using an automatic system for the irrigation in each line, according to the water content in the soil. This system allows also the data acquisition, in order to calculate, day by day, the water consumption for each line covered with different mulch films. According to the experimental results (see Fig. 1.13) it has been demonstrated that the use of photo-reflective mulch film could give a valid contribution to solving the water problem in agriculture.

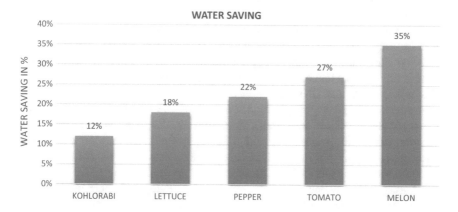

Fig. 1.13 Evidence of saving water in agriculture for some vegetables

1.11   Quality Control and European Rules as References

All companies producing plastic films for agriculture are responsible for ensuring the technical specifications of their films, and performing quality control both on the characteristics of raw materials and output product. Samples are randomly taken directly from the production, and sent to the lab for chemical–physical analysis designed to test the technical specifications of each product in order to certify the declared quality standards. In the laboratory, taken into account the quality standards imposed by the international regulation, samples are tested in order to verify the optical characteristics, mechanical and thermal properties of different production lots.

The standard tests on the selected products are:

- percentage elongation at break
- tensile strength
- tear strength
- impact strength
- measurement of film thickness precision
- transmission measurement in the region of UV-Vis-NIR
- measurement of thermal insulation
- anti-drip system.

First of all, always with the random method, tests are performed on the incoming raw material to verify if the characteristics of the polymers and additives are those declared in the contract with the supplier of the raw material.

The standard tests are:

- measurement of the melt flow rate
- thermogravimetric analysis
- humidity test.

At the end of the path of the Quality Control, products, once pass all tests, are guaranteed and certified according to the relevant legislation. As a reference, it is reported here an abstract of the European Regulation UNI EN 13206.

The main objective of the European Regulation EN 13206 is to categorize thermoplastic cover films, used in agriculture and horticulture, based on their expected life time. This new European regulation uses rational criteria, able to provide reliable conclusions about the expected lifespan of a cover film during use. This rule replaces the UNI 9298 establishing the features and requirements that must possess certificates for plastic films covering the various tunnels and greenhouses (small, medium and large). It establishes the physical and mechanical requirements of the transparent and diffusing polyethylene and/or copolymers plastic films, which are intended to cover greenhouses. The standard takes into account materials like LDPE (low density polyethylene), LLDPE (linear low-density polyethylene), EVA (Ethyl Vinyl acetate) and their mixtures and divides the films into three categories:

- regular film with good total transmittance (88–86%) and low greenhouse effect;
- thermal film with high total transmittance (89–87%) and elevated greenhouse effect (55–75%);
- film diffused light with less total transmittance (85–80%) and elevated greenhouse (60–75%).

A final consideration

A shrewd choice of plastic films, on the base of each farmer's need, has to take into account the following characteristics:

(1) *total transmittance to solar radiation*, such as light quantity crossing through the film and that gives rise to the greenhouse effect, measured by spectrophotometer in the field of far infrared radiation (wavelength between 7000 and 20,000 nm) held in the greenhouse. This feature is reported in percentage (%). In the case of thermal films, it is important to have also the value of thermicity, expressed in percentage;

(2) *mechanical properties*, such as tensile strength, tear strength, impact strength, percentage elongation at break;

(3) *service life*, established on the basis of percentage elongation at break of the original film, must be greater than 50% as determined by ISO 4892-2; the materials are classified into several classes, distinguished by the letters N, A, B, C, D and E, corresponding to films for seasonal life, yearly life, long life, with regards to conditions of luminous intensity;

(4) *respect for the environment*, according to which the material once ceased its operation, should be collected and recycled, without causing harm to the environment;

(5) *uniform thickness and width*, with a tolerance of ±5%.

References

1. Emmert EM (1955) Low-cost Plastic greenhouses (Progress report/Kentucky Agricultural Experiment Station)
2. Garnaud JC (2000) A milestone for a history of progress in plasticulture. Plasticulture 119:30–43
3. Jouet JP (2001) Plastics in the world. Plasticulture 120:108–126
4. Jouet JP (2004) The situation of plasticulture in the world. Plasticulture 123:48–57
5. Jiang W, Qu D (2004) Mu D. and Wang L. R., China's energy-saving greenhouses. Chronica Hortic 44(1):15–17
6. Reynolds A (2009) Agricultural film markets, trends and business development. Agricultural film 2009. In: The international conference for agricultural & horticultural film industry, Applied Market Information Ltd., Barcelona, 24–26 Feb 2009

7. Reynolds A (2010) Updated views on the development opportunities in agricultural film markets. Agricultural film 2010. In: International conference on greenhouse, tunnel, mulch and agricultural films and covers. Applied Market Information Ltd., Barcelona, 22–24 Nov 2010

8. Garnaud JC (1988) Agricultural and horticultural application of polymers. Rapra Technology Ltd., Pergamon Press, Oxford

9. Brown RP (2004) Polymers in agriculture and horticulture. Rapra Technology Ltd., UK

10. AMI (2014) Agricultural Film 2014. In: International industry conference on silage, mulch greenhouse and tunnel films used in agriculture, Barcelona, Sep 2014

11. Edser C (2002) Light manipulating addutives extend opportunities for agricultural plastic films. Plast Addit Compd 4(3):20–24

12. Waaijenberg D, Gbiorczyk K, Feuilloley P, Verlodt I, Bonora M (2004) Measurement of optical properties of greenhouse cladding materials; harmonisation and standardisation. Acta Hortic (ISHS) 633:107–113

13. Gbiorczyk K, von Elsner B, Sonneveld PJ, Bot GPA (2002) The effect of roof inclination on the condensation behaviour of plastic films used as greenhouse covering materials. Acta Hortic (ISHS) 633:127–136

14. Espi E, Salmeron A, Fontecha A, Garcia A, Real AI (2006) Plastic films for agricultural application. J Plast Film Sheeting 22(85):85–102

15. Espi E, Salmeron A, Fontecha A, Garcia A, Real AI (2006) New ultrathermic films for greenhouse cover. J Plast Film Sheeting 22(1):59–68

16. Ramakrishna A et al (2006) Effect of mulch on soil temperature, moisture, weed infestation and yield of groundnut in northern Vietnam. Field Crops Res 95:115–125

17. Adam A, Kouider SA, Hamou A, Saiter JA (2005) Studies of polyethylene multi layer films used as greenhouse covers under Saharan climatic conditions. Polym Testing 24(7):834–838

18. Papadakis G, Briassoulis D, Scarascia Mugnozza G, Vox G, Feuilloley P, Stoffers JA (2000) Review Paper (SE- Structures and Environmental): Radiometric and thermal properties of, and testing methods for, greenhouse covering materials, J Agric Eng Res 77(1):7–38

19. Schreiner M, Mewis I, Huyskens-Keil S, Jansen MAK, Zrenner R, Winkler JB, O'Brien N, Krumbein A (2012) UV-B induced secondary plant metabolites—potential benefits for plant and human heath. Crit Rev Plant Sci 31:229–240

20. Katan J, DeVay JE (1991) Soil solarization. CRC Press, Boca Raton

21. Krueger R, McSorley R (2009) Solarization for pest management in Florida. ENY-902, Entomology and Nematology Department, Florida Cooperative Extension Service, IFAS, University of Florida, Gainesville, FL. http://edis.ifas.ufl.edu/in824

22. Mormile P, Capasso R, Rippa M, Petti L (2013) Light filtering by innovative plastic films for mulching and soil solarization. Acta Hortic 1015:113–121

23. Mormile P, Rippa M, Petti L (2013) A combined system for a more efficient soil solarization. Plasticulture 10(132):44–55

24. Mormile P, Immirzi B, Lahoz E Malinconico M, Morra L, Petti L, Rippa M (2015) Improvement of soil solarization through a hybrid system simulating a solar hot water panel, ICSEA 2015, NY City

25. Fennimore S, Ajwa H (2011) Totally impermeable film retains fumigants, allowing lower application rates in strawberry. Calif Agric 65(4):211–215

26. Bangarwa SK, Norsworthy JK, Gbur EE, Mattice JD (2010) Phenyl isothiocyanate performance on purple nutsedge under virtually impermeable film mulch. HortTechnology 20(2):402–408

27. Cerny T, Faust J, Layne D, Rajapakse N (2003) Influence of photo-selective films and growing season on stem growth and flowering of six plant species. J Amer Soc Hort Sci 128:486–491

28. Schettini E, De Salvador FR, Scarascia Mugnozza G, Vox G (2011) Radiometric properties of photo-selective and photoluminescent greenhouse plastic films and their effects on peach and cherry tree growth. J Hortic Sci Biotechnol 86(1):79–83
29. Mormile P, Rippa M, Bonanomi G, Scala F, Yan C, Petti L (2015) Photo-reflective mulches for saving water in agriculture, world academy of science, engineering and technology. Agric Biosyst Eng 2(5)

Chapter 2
Biodegradable and Biobased Plastics: An Overview

Ramani Narayan

Abstract Plastic mulch film and sheets, rods, and tubing find increasing use in agriculture. Current polyethylene plastic mulch film is not biodegradable and therefore cannot be plowed back into the soil. It may undergo fragmentation, and the small fragments are blown all over and find its way into ocean and other pristine environments. This causes irreparable harm to ecosystems and the habitats. Completely soil-biodegradable plastics or compostable plastics offer an environmentally responsible end-of-life solution for plastic mulch film and plasticulture products. Claims of biodegradability should be qualified by the disposal environment (soil or compost), 90% + biodegradability as measured by the evolved CO_2 from the microbial process using international standards for soil biodegradability and/or compostability. However, one has to be careful of misleading claims that are prevalent in the marketplace, particularly additive-based polyolefin plastics. Using biobased carbon in place of petro-fossil carbon in the products offers a reduced carbon footprint, empowers rural agrarian economy, and reduces dependence on fossil resources.

Keywords Plasticulture · Biodegradability · Compostability · Biobased carbon · Biobased plastics

Plastic mulch film is standard practice used in agriculture to control weeds, increase crop yield, and shorten time to harvest. Other plastic materials like sheets, rods, and tubing find use in agriculture like greenhouses, small tunnels, fruit and vegetable coverings, and many more applications. About 5 million tons of plastic resin is used worldwide and growing. It contributes significantly to the economic viability of farmers, providing increased harvests, less reliance on herbicides and pesticides, more efficient water conservation, reduced plant disease, and better protection of food products. However, disposal is a major issue. The polyethylene plastic mulch film is not biodegradable in the soil environment; therefore, it cannot be plowed

R. Narayan (✉)
Michigan State University, East Lansing, MI 48824, USA
e-mail: narayan@msu.edu

© Springer-Verlag GmbH Germany 2017
M. Malinconico (ed.), *Soil Degradable Bioplastics for a Sustainable Modern Agriculture*, Green Chemistry and Sustainable Technology,
DOI 10.1007/978-3-662-54130-2_2

23

back into the soil. It may undergo fragmentation, and the small fragment is blown all over and finds its way into ocean and other pristine environments.

Burning these plastic films in the fields is not an option because it contributes to serious environmental air and particulate pollution. Therefore, this practice is being phased out and many countries have strict regulatory bans on plastic film burning. Collection, cleaning, and recycling these plastics to same or other products offers an approach to managing plastic mulch film waste, and there are several companies that offer these services.

Designing for complete biodegradability in soil after its intended use as mulch film or other plasticulture products offers an environmentally responsible value proposition. Alternatively, these mulch films can be collected and composted or anaerobically digested with farm and other bio wastes. The "complete biodegradability" approach has the potential to be more economically and technically viable than recycling as there is no need to transport, clean, recycle to usable product, and finally find markets for these recycled products.

Unfortunately, "biodegradability" is a much misused term and many misleading claims abound in the marketplace. This chapter provides a fundamental understanding of "biodegradability" and explains the science behind it. International standards to measure biodegradability in several environments like composting, anaerobic digestion, and soil are presented. Learning to understand and recognize misleading claims is discussed. This chapter also introduces the concept of "biobased" and its attendant value proposition. Biobased products offer the value proposition of a reduced carbon footprint, empowered rural agrarian economy, and reduced fossil resources dependence [1, 2].

2.1   Biodegradability—The Science

Biodegradability is an end-of-life option that allows one to harness the power of microorganism present in the selected disposal environment to completely remove plastic products designed for biodegradability from the environmental compartment via the microbial food chain in a timely, safe, and efficacious manner. Terms like "oxo", "hydro", "chemo", "photo" degradable describe abiotic (nonbiological process) mechanisms of degradation. They do not constitute or represent "biodegradability"—the biological process by which microorganisms present in the disposal environment assimilate/utilize carbon substrates as food for their life processes—see Fig. 2.1.

Because it is an end-of-life option, and harnesses microorganisms present in the selected disposal environment, one must clearly identify the "disposal environment" when discussing or reporting the biodegradability of a product—for example, biodegradability in composting environment (compostable plastic), biodegradability in soil environment, biodegradability under anaerobic conditions (in anaerobic digester environment or even a landfill environment), or biodegradability in marine environment.

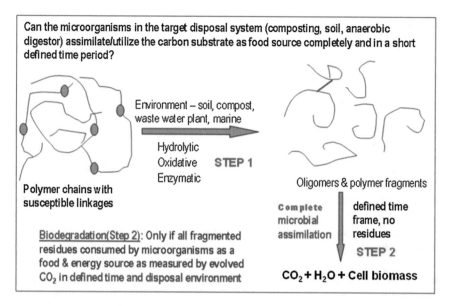

Can the microorganisms in the target disposal system (composting, soil, anaerobic digestor) assimilate/utilize the carbon substrate as food source completely and in a short defined time period?

Environment – soil, compost, waste water plant, marine

Hydrolytic
Oxidative **STEP 1**
Enzymatic

Polymer chains with susceptible linkages

Oligomers & polymer fragments

Biodegradation(Step 2): Only if all fragmented residues consumed by microorganisms as a food & energy source as measured by evolved CO_2 in defined time and disposal environment

Complete microbial assimilation | defined time frame, no residues

STEP 2

$$CO_2 + H_2O + Cell\ biomass$$

Fig. 2.1 Understanding the biodegradability process

Reporting time to complete biodegradation or more specifically, the time required for the complete microbial assimilation of the plastic in the selected disposal environment is an important requirement. Product claims stating that plastic will eventually biodegrade based on data showing an initial 10–20% biodegradability is not acceptable and very misleading especially since the percent biodegradation levels off and reaches a plateau after some initial rate and level of biodegradation. Drawing a straight line extrapolation from the initial rate and value to 100% biodegradation is scientifically untenable, and unfortunately many of the claims are based on this type of extrapolation.

2.2 Measuring and Reporting Biodegradability

Basic biology teaches how microorganisms utilize carbon substrates as food for their life processes. Carbon substrates including any biodegradable plastics have to be transported into the microbial cell. The transport is governed by several factors like molecular weight, structure, functional groups, hydrophilic-hydrophobic balance, and other special factors. Inside the cell, the carbon is biologically oxidized to CO_2 releasing energy that is harnessed by the microorganisms for its life processes. Under anaerobic conditions, $CO_2 + CH_4$ (biogas) are produced (see Fig. 2.2). Thus, a measure of the rate and amount of CO_2 or $CO_2 + CH_4$ evolved as a function of total carbon input to the process is a direct measure of the amount of carbon substrate being utilized by the microorganism (percent biodegradation).

Aerobic process

Glucose/C-bioplastic + 6O$_2$ $\longrightarrow$ **6CO$_2\uparrow$ + 6H$_2$O; $\Delta G^{0'}$= -686 kcal/mol**

Anaerobic process

Glucose/C-bioplastic $\longrightarrow$ **2 lactate** $\longrightarrow$ **3CO$_2\uparrow$+ 3CH$_4\uparrow$ $\Delta G^{0'}$= -47 kcal/mol**

Fig. 2.2 Basics for the microbial utilization of carbon substrates

It would seem obvious and logical from the above basic biology lesson that to make a **claim of biodegradability**, all that one needs to do is the following: Expose the test plastic substrate as the sole carbon source to microorganisms present in the target disposal environment (like composting, or soil or anaerobic digestion or marine), and measure the CO$_2$ (aerobic) or CO$_2$ + CH$_4$ (anaerobic) evolved. A measure of the evolved gas provides a direct measure of the plastics carbon being utilized by the microorganisms present in the target disposal environment (% biodegradation). ASTM, EN, and ISO test methods teach how to measure the percent biodegradability in different disposal environments based on the fundamental biochemistry described above—irrespective of what the initial degradation is—oxo, hydro, chemo—the abiotic degradation.

Thus, one can measure the rate and extent of biodegradation or microbial utilization of the test plastic material using it as the sole carbon source in a test system containing a microbial rich matrix like compost or soil, in the presence of air and under optimal temperature conditions (preferably at 58 °C—representing the thermophilic phase). Figure 2.3 shows a typical graphical output that would be obtained if one were to plot the percent carbon converted to CO$_2$ as a function of time in days. First, a lag phase during which the microbial population adapts to the available test C-substrate. Then, the biodegradation phase during which the adapted microbial population begins to utilize the carbon substrate for its cellular life processes, as measured by the conversion of the carbon in the test material to CO$_2$. Finally, the output reaches a plateau when all of the substrate is completely utilized. Linear or any other form of data extrapolation from these complex biological systems is not acceptable and is very misleading because credible scientific substantiation for the extrapolation model does not exist.

Claims of degradable, partial or extrapolated biodegradability or eventual biodegradable are not acceptable, because it has been shown that these degraded fragments absorb toxins present in the environment, concentrating them and transporting them up the food chain [4]. Therefore, complete removal from the disposal environment in a short time period of 1–2 years is essential to eliminate potentially serious human health and environmental consequences.

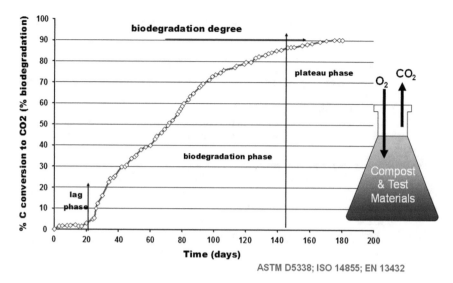

Fig. 2.3 Measuring rate and extent of biodegradability using text plastic as the sole carbon source (Reproduced with permission from [3] copyright @ 2012, American Chemical Society)

2.3   International Standards for Biodegradability

To meet the requirements of biodegradability under industrial composting conditions (compostable plastics), a plastic must satisfy the primary requirement of complete microbial utilization (biodegradability) as measured by the evolved CO_2 under composting or soil environment, as discussed in the earlier section. In addition, it has to meet the disintegration and safety criteria to claim compostability. ASTM D6400, D6868, ISO 17055, and EN 13432 are specification standards for compostable plastics. Another ISO specification standard ISO 18606 addresses "Packaging and the environment—Organic recycling" and follows the same basic principle outlined above.

Specification standards set the pass/fail criteria and is based on a standard test method. Standard test methods teach practitioners how to conduct biodegradability tests in the selected environment, how to collect data from the test, and how to correctly report the results of the tests. There are no pass/fail criteria. Unfortunately, many companies make unqualified claim of biodegradability and label their plastic product "biodegradable" referencing a standard test method without providing actual percent biodegradability values obtained in the test—a graphical display of the percent biodegradability (measured by the evolved CO_2 or $CO_2 + CH_4$) as a function of time in days. This is misleading as the consumer or stakeholder assumes that the product is completely biodegradable in a short time period. Tables 2.1 and 2.2 provide a list of international specification standards and test methods.

Table 2.1 List of international specification standards that have specified pass/fail criteria in the target disposal environment

• **Biodegradability under composting conditions—Compostable Bioplastics**
        • **ASTM D6400**—Specification for compostable plastics
        • **ASTM D6868**—Specification for plastics coatings and modifiers on paper and other compostable substrates
        • **ISO 17088**—Specification for compostable plastics
        • **EN 13432**—Specification for compostable packaging—focus on packaging
        • **ISO 18606**—Packaging and the Environment—Organic recycling—focus on packaging

• **Biodegradability under marine environment**   • **D7021**—Specification for nonfloating biodegradable plastics in the marine environment
• **Biodegradability under soil environment**   • **ASTM**—under development—90% carbon assimilation by microorganism as measured by evolved CO_2 in 2 years or less using ASTM D5988

Table 2.2 List of biodegradability test method standards from ASTM and ISO

• **ASTM D5338**—test method for measuring biodegradability under composting environment
• **ASTM D5988**—test method for measuring biodegradability in soil environment
• **ASTM D5511**—test method for measuring biodegradability in a high solids anaerobic digestor
• **ASTM D5526**—test method for measuring biodegradability in a landfill/bioreactor environment
• *ISO 14852—ultimate **aerobic biodegradability** of plastic materials in an aqueous medium—Method by analysis of evolved carbon dioxide*
• *ISO 14853—ultimate **anaerobic biodegradability** in an aqueous system—Method by measurement of biogas production*
• *ISO 14855—Determination of the ultimate **aerobic biodegradability** of plastic materials under **controlled composting conditions**—Part 1: Method by analysis of evolved carbon dioxide and Part 2: Gravimetric measurement of carbon dioxide evolved in a laboratory-scale test*
• NOTE—Standard test methods teach how to conduct the test and report the results. It has no pass/fail criteria and should not be used to make broad claims of biodegradability. Reporting should strictly follow the procedures laid out in the test methods showing the disposal environment, percent biodegradation, and time to achieve that biodegradation. Extrapolation of data is not permitted in these test methods

2.4   Misleading Claims of Biodegradability

There are additive-based plastics—oxo-degradable and organic additives added at 1–2% levels to conventional hydrocarbon resins like polyethylenes (PE), polypropylene (PP), polystyrene (PS), polyethylene terephthalate (PET), and other plastics that are claimed to make them "biodegradable". However, the fundamental biological data showing percent carbon utilized or assimilated by the microorganisms, as measured by the evolved CO_2 (aerobic) or $CO_2 + CH_4$ (anaerobic), are not provided. Some of the data show 10–20% biodegradation which then levels off with little or no biodegradation. Weight loss, molecular weight reductions, carbonyl index, mechanical property loss, biofilm formation, and microbial colonization do not confirm the microbial utilization of the polymeric carbon substrate, nor do they provide the amount of carbon utilized or the time to complete microbial utilization.

2.5 U.S. Federal Trade Commission (FTC) Green Guides [5]

The U.S. Federal Trade Commission (FTC) recently issued new Green Guides, on Environmental Marketing Claims to help marketers avoid deceptive environmental claims. The FTC guides state that an "unqualified degradable claim for items entering the solid waste stream should be substantiated with competent and reliable scientific evidence that the entire item will **fully decompose** (break down and return to nature; i.e., decompose into elements found in nature) **within one year** after customary disposal." It also emphasizes that unqualified degradable/biodegradable claims for items that are customarily disposed in landfills, incinerators, and recycling facilities are deceptive because these locations do not present conditions in which complete decomposition will occur within one year.

The term fully decompose into elements found in nature equates to the complete abiotic and biotic breakdown of the plastic to CO_2, water, and cell biomass via microbial metabolism. This was discussed in detail in the earlier sections.

Degradable claims can be made if it is qualified clearly and prominently to the extent necessary to avoid deception about:

- The product's or package's ability to degrade in the environment where it is customarily disposed and more importantly **the rate and extent of degradation/biodegradation**.

In the case of biodegradability claims, one has to provide "reliable and competent science based evidence" of the rate and extent of biodegradation in the target disposal environment – a graphical plot of percent biodegradability as measured by the evolved CO_2 (aerobic) or $CO_2 + CH_4$ (anaerobic) versus time in days. The FTC guides do not identify any specific testing protocol or specification and therefore reserve the right to evaluate the data which forms the basis of the claims. However, they clearly require that the evidence should be based on standards generally accepted in the relevant scientific fields. So ASTM, EN, ISO standards should be used to provide the evidence for validating the rate and extent of biodegradation in the selected disposal environment(s).

In summary, claims of a plastic product's biodegradability must be qualified by graphically showing the percent product carbon being utilized (percent biodegradation) by microorganisms present in the selected disposal environment as measured by the evolved CO_2 (aerobic) or $CO_2 + CH_4$ (anaerobic) as a function of time in days in the selected disposal environment.

2.6 Biobased Plastics

Biobased plastics are "plastics in which the (organic) carbon (of the polymer molecule) in part or whole comes from plant-biomass like agricultural crops and residues, marine and forestry materials, algae, and fungi **living in a natural environment in equilibrium with the atmosphere**". Figure 2.4 explains the fundamental concept behind biobased plastics

Note: Plastics in which the (organic) carbon comes from petroleum, natural gas, and other fossil resources are not biobased.

Plastics—Material which contains as an essential ingredient a carbon-based high polymer and which at some stages in its processing into finished product can be shaped by flow

Biobased—containing organic carbon of renewable origin like (from) agricultural, plant, animal, fungi, microorganisms, marine, or forestry materials living in a natural environment in equilibrium with the atmosphere—ASTM D6866

Organic Material(s)—material(s) containing carbon-based compound(s) in which the carbon is attached to other carbon atom(s), hydrogen, oxygen, or other elements in a chain, ring, or three dimensional structures—IUPAC nomenclature

ASTM D6866-16—Standard Test Methods for Determining the Biobased Content of Solid, Liquid, and Gaseous Samples Using Radiocarbon Analysis provide the following definitions related to biobased plastics:

Biobased—containing organic carbon of renewable origin like agricultural, plant, animal, fungi, microorganisms, marine, or forestry materials living in a natural environment in equilibrium with the atmosphere.

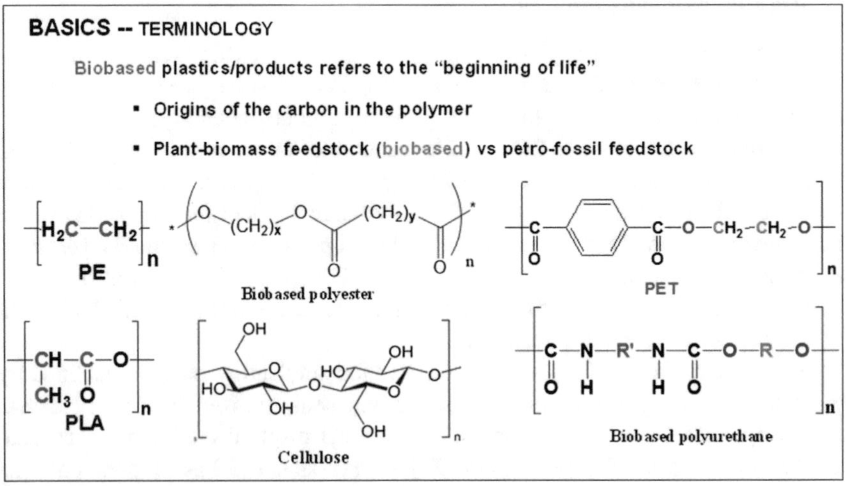

Fig. 2.4 Fundamental concepts for biobased plastics

biobased carbon content—the amount of biobased carbon in the material or product as a percent of the total organic carbon (TOC) in the product

biobased carbon content on mass basis—amount of biobased carbon in the material or product as a percent of the total mass of product

biogenic—containing carbon (organic and inorganic) of renewable origin like agricultural, plant, animal, fungi, microorganisms, macroorganisms, marine, or forestry materials

biogenic carbon content—the amount of biobased carbon in the material or product as a percent of the total carbon (TC) in the product

biogenic carbon content on mass basis—amount of biogenic carbon in the material or product as a percent of the total mass of product.

2.7   Application

USDA biopreferred program and EPA Greenhouse gas reporting requirements (D7459) use ASTM D6866 as does Japan EcoMark (http://www.ecomark.jp) program. The EU-CEN standards are in harmony with ASTM and ISO standards and use the same basic principles of radiocarbon analysis enunciated in ASTM D6866. European certification organizations are Vincotte, Belgium (OK biobased), DIN-CERTCO (Germany).

The biobased carbon value proposition for plastics does not address its end-of-life—the question of what happens to product after use when it enters the disposal environment. Biobased plastics are not necessarily biodegradable-compostable and all biodegradable-compostable plastics are not automatically biobased. The biobased carbon content has zero impact on the end-of-life of the biodegradable plastics. The molecular structure of the plastic and the availability of its carbon for transport into the microbial cell and subsequent utilization for energy drive the microbial assimilation (percent biodegradability) of carbon substrates like plastics—the availability of carbon in a molecule to the microbes and not the source of the carbon is the key learning.

Value proposition for "biobased"

Replacing petro/fossil carbon with biobased carbon (from plant-biomass feed-stocks) in plastics and industrial products offers the value proposition of removing carbon present as CO_2 in the environment and incorporating it into a polymer molecule via plant-biomass photosynthesis in a short time scale of 1 (agricultural crops, algae) to 10 years (short rotation wood and tree plantations) in harmony with Nature's biological carbon cycle. Plastics made from petro/fossil resources (like Oil, Coal, Natural gas) which are formed from plant-biomass over millions of years and so cannot be credited with any CO_2 removal from the environment even over a hundred-year time scale (the time period used in measuring global warming

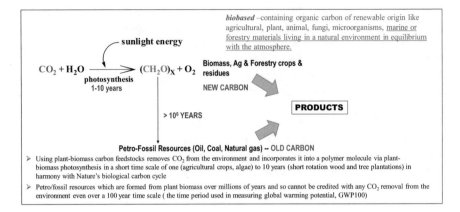

Fig. 2.7 Illustrating zero material carbon footprint using biobased carbon

potential, GWP100). Process carbon and environmental footprint (arising from the process of converting the feedstock to product) are also improved. This concept is shown in Fig. 2.5.

The biobased carbon content of products is determined independently and unequivocally using radio carbon analysis as codified in international standards—the primary one is the ASTM D6866 (Standard Test Method for determining biobased (carbon) content of solids, liquids, and gaseous samples using radiocarbon analysis). Using experimentally determined biobased carbon content and applying fundamental stoichiometric calculations, one can readily calculate the amount of CO_2 removed from the environment by 1 kg of material. For example: 1 kg of biobased polyethylene (PE) containing 100% biobased carbon content would result in removing 3.14 kg of CO_2 from the environment. 1 kg of PLA (100% biobased carbon content) would remove 1.83 kg of CO_2 from the environment. 1 kg of the current bio PET (20% biobased carbon content—only the glycol carbons come from plant-biomass) results in 0.46 kg of CO_2 removal from the environment. 1 kg of the 100% biobased carbon content PET results in 2.29 kg of CO_2 removal. In contrast, the petro-fossil carbon-based products result in zero CO_2 removal from the environment. These results are graphically shown in Fig. 2.6.

Eventually, at the end-of-life of these plastics, the carbon will be released back into the environment as CO_2 through waste-to-energy systems or incineration or through composting or anaerobic digestion (if it has biodegradability-compostability feature built into it). However, the CO_2 released will be captured by the next season's crop or biomass plantation resulting in a net zero material carbon footprint, in harmony with Nature's carbon cycle. In contrast, the non-biobased PE or PP will contribute a net 3.14 kg of CO_2 into the environment for every 1 kg of PE used. 1 kg of PET will contribute 2.29 kg of CO_2 to the environment. Figure 2.7 graphically reports these numbers and illustrates the zero carbon footprint concept.

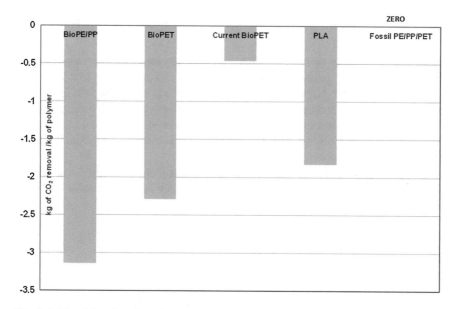

Fig. 2.6 Material carbon footprint—Amount of CO_2 removed from the environment per kg of resin

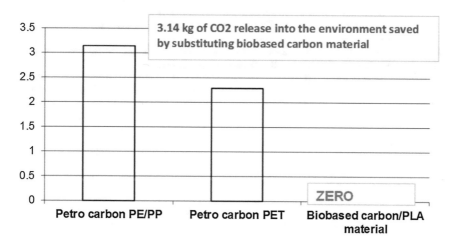

Fig. 2.5 Value Proposition for using biobased versus petro/fossil carbon—the material carbon footprint

In summary, the replacement of petro-fossil carbon in whole or part by biobased carbon (derived from plant-biomass resources) offers the value proposition of reduced carbon footprint and the enabling technology to move toward the closed loop "circular economy" model that is being advocated and adopted by many nations and major industrial organizations and brand owners.

2.8 Conclusions

Truly and completely soil-biodegradable plastics or compostable plastics offer an environmentally responsible end-of-life solution for plastic mulch film and other plasticulture products. However, one has to be careful of misleading claims that are prevalent in the marketplace, especially additive-based polyolefin plastics. International standards for soil biodegradability or compostability should be met for claims of biodegradability and the disposal environment, extent and rate of biodegradation should be clearly documented.

Using biobased carbon in place of petro-fossil carbon in the products offers a value proposition of reduced carbon footprint, empowered rural agrarian economy, and reduced fossil resource dependence.

References

1. Narayan R (2011) Carbon footprint of bioplastics using biocarbon content analysis and life cycle assessment, MRS (Materials Research Society) Bull 36(09):716–721
2. Narayan R (2014) Principles, drivers, and analysis of Biodegradable and Biobased Plastics (Chap. 16). In: Bastioli C (ed) Handbook of biodegradable polymers, 2nd edn, Smithers RapraTechnology, Nov 2014, ISBN-13:978-1847355270 ISBN-10: 1847355277
3. Narayan R (2012) Biobased & biodegradable polymer materials: rationale, drivers, and technology exemplars (Chap. 2). ACS (an American Chemical Society publication) Symp Ser 1114:13–31
4. Thompson RC, Moore CJ, vom Saal FS, Swan SH (2009) Plastics, the environment and human health. Phil Trans R Soc London Ser B 364(1526):2127–2139
5. Federal Register/vol 77(197)/2012/ Rules and Regulations; FEDERAL TRADE COMMISSION 16 CFR Part 260 Guides for the Use of Environmental Marketing Claims

Chapter 3
Biodegradable Materials in Agriculture: Case Histories and Perspectives

Sara Guerrini, Giorgio Borreani and Henk Voojis

Abstract Many applications of traditional plastic in agriculture have a short life, which on average do not exceed 2 years and for precisely this characteristic generates large quantities of waste that must be adequately disposed. It has been estimated that 30% of the plastic waste produced from agriculture originates from short-life applications such as clips, wires, nets, pheromone dispensers, and geotextiles with a high risk of remaining in the agricultural system thereby causing pollution. In this chapter the state of the art of these commercial applications is described, together with research and development for these applications. Some successful case histories in European countries are reported, along with how these solutions came about. Biodegradable and compostable bioplastics have physicochemical and mechanical characteristics suitable to substitute traditional polymer plastic applications, thereby reducing waste generation at the end of life cycle. Biodegradable and compostable materials, compliant with the main international standards can be left directly in the field on the soil where they are biodegraded by microorganisms present either in the soil or in the compost heaps, together with crop residues, producing organic matter that can be recycled to the soil.

Keywords Biodegradable and compostable materials · Bioplastics · Compostable clips · Compostable pheromone dispenser · Compostable geotextile · Compostable silage films

S. Guerrini (✉)
Novamont, Novara, Italy
e-mail: Sara.Guerrini@novamont.com

G. Borreani
DISAFA, University of Turin, Turin, Italy

H. Voojis
Amsterdam, The Netherlands

© Springer-Verlag GmbH Germany 2017
M. Malinconico (ed.), *Soil Degradable Bioplastics for a Sustainable Modern Agriculture*, Green Chemistry and Sustainable Technology,
DOI 10.1007/978-3-662-54130-2_3

3.1 State of the Art of Different Applications of Plastic in Agriculture

There is a wide range of plastic tools used in agriculture; some of these in recent years using biodegradable materials have been implemented with success. However, still many of these could be efficiently substituted by optimizing techniques and performance of these low environmental impact materials. In the following pages some innovative applications of biodegradable materials which, at different levels, can be found at pilot stage, or commercially available will be described. The aim is to give as much as possible, the state of the art of the still increasing possibility of biodegradable and compostable materials in agriculture to reduce the production of plastic waste and the costs connected to the disposal and end of life of traditional plastic materials, offering different solutions.

In Table 3.1 the main applications of plastic materials in various agriculture sectors are reported: from greenhouse films to protection nets and irrigation drip tubes. Plastic is indeed an important material for agriculture: increasing yield and quality of crops, and helping in the reduction of chemical inputs and water in the agro-systems [1].

It is possible to indicate a series of characteristics that are common to various applications which make them easier to be partially or completely substituted by biodegradable and compostable plastics:

– short-medium shelf life in the field: on average from one to three seasons (3 years);
– plastic waste generated at the end of use: not toxic, and therefore no further cleaning operations are required before collection (e.g., on the contrary plastic containers or cans for agrochemicals need cleaning);
– plastic waste generated at the end of use is highly contaminated with soil or other plant residues (e.g., nets for growing crops under greenhouses, geotextile used in reforestation, or in landscaping) and therefore recycling can be difficult and in several cases uneconomical;

Table 3.1 Main applications of plastic materials in the agriculture sector

Protected cultivation films:	Nets:	Packaging:
• Greenhouse and tunnel	• Anti-hailstone	• Fertilizer sacks
• Low tunnel	• Anti-bird	• Agrochemical cans
• Mulching	• Wind breaking	• Containers
• Nursery film	• Shading	• Tanks for liquid storage
• Direct covering	• Harvesting	• Crates
• Covering vineyards and orchards	**Piping, irrigation/drainage**:	**Other:**
	• Water reservoir	• Silage films
	• Channel lining	• Fumigation films
	• Irrigation tapes and pipes	• Balet wine
	• Drainage pipes	• Bale wraps
	• Micro-irrigation	• Nursery pots
	• Drippers	• Strings and ropes

- plastic waste difficult to be separated from plant residues or collected (e.g., pheromone dispensers, clips, and twines). In this case the organic residue is contaminated with plastic and cannot be used for composting;
- waste management: most of the time the end of life of this type of plastic is the incineration with or without energy recovery.

For these types of tools biodegradable polymers can be a solution to reduce the overall environmental impact of some applications and increase the quantity and quality of organic material recuperated via composting (green waste).

In Europe and around the world, waste prevention and using waste as a resource is becoming more and more important, not only in environmental policies, but also as a key aspect of the transition toward a green economy. In Europe, from the Roadmap to a Resource Efficient Europe and its ambition plan to reduce waste generation per person (EC, 2011), to the European Commission Communication "towards a circular economy," ("A zero waste programme for Europe," EC, 2014) this direction has been clearly underlined. The need of product design approaches intended to reduce the quantity of material required to deliver a particular service as well as the use of raw materials which hazardous or difficult to recycle (both products and production processes) are aspects that need to be taken in account also for the agricultural sector. Biodegradable and compostable materials represent possible important solutions to reduce the plastic waste at the end of many processes and crop cultivations.

In the following pages some case histories of applications of biodegradable materials at commercial level in different sectors will be described:

- *The greenhouse*: use of biodegradable clips and twines in Dutch greenhouses;
- *The orchard*: use of biodegradable pheromone dispensers in Italian orchards;
- *The forest and public green areas*: the potential of biodegradable and compostable geotextiles in avoiding the dispersion of long-life mulch films in the natural environment.

Furthermore, a view of developing applications of biodegradable materials in important sectors such as silage protection films and irrigation drippers will be given.

For each case history and application, a market overview of the present in terms of quantity and characteristics of plastic applications will be presented, together with the context and the main reasons which led to the substitution with a biodegradable and compostable material. It will be also underlined the important role of legislation and standards as drivers of the innovation processes.

The market data reported are intended to indicate a possible alternative model for agriculture and landscaping sectors in order to both reduce the end of life production of plastic waste and to gain organic matter via biological transformation of "cleaner" green waste (compost) which can be reintroduced into the soil. At the same time biodegradable materials can help to solve the problem of plastic pollution in the environment, thanks to their intrinsic biodegradability.

3.1.1 Market Data on Plastic Applications in Agriculture

It is not an easy task to find reliable data on the world consumption of plastic materials in the agricultural sectors when our need is to focus on the different single applications. Nevertheless, a good starting point is the data gathered by Jouët for the CIPA congress in 2006 and presented in Table 3.2. Data on plastic tools are reported as hectares covered by the applications. From 1985 to 2005 there was a constant increase in the use of plastic materials for agriculture, and this world trend is continuing [2].

Nevertheless some areas are growing less than others. For example Europe, where the market is already mature, it can be considered at a steady state. On the other end, Asia (China) and South America are now fast growing markets [3].

In Europe, where the plastic agriculture market is mature both for use and end of life management, a datum appears quite evident: each year about "only" about 300,000 tons of plastic waste are collected from this sector. The waste comes mainly from used plastic films, which can be collected, stocked, and recuperated in different ways, but only about 50% of the total plastic waste is recuperated after use and properly disposed (Table 3.3) [4].

Agricultural plastic waste which is not properly recovered and left in the ground (landfilling estimated values: 0.67 Mtons/year in Europe) or burnt uncontrollably (estimated value: 0.37 Mtons/year in Europe) produces the release in the environment of harmful substances (e.g., 12% of dioxin or furan emissions come from the agriculture sector) [5].

However, both landfilling and incineration do not seem to be the best practices to dispose of plastic waste. Landfilling is expensive, an environmentally less feasible

Table 3.2 World consumption of plastic for agriculture applications from 1985 to 2005 (metric tons)

Applications	1985	1991	1999	2002	2005
Low tunnels	88,000	122,000	168,000	170,000	178,000
Mulching	270,000	370,000	650,000	670,000	730,300
Direct covers	22,500	27,000	40,000	48,000	48,000
Greenhouses and large tunnels	180,000	350,000	450,000	560,000	560,000
Silage films	140,000	265,000	540,000	560,000	560,000
Polypropylene (PP) twine	100,000	140,000	204,000	195,000	190,000
Hydroponic systems	5000	10,000	20,000	25,000	26,000
Micro-irrigation	260,000	325,000	625,000	720,000	920,000
Others: nets, plastic bags, except packaging for fertilizers	80,000	130,000	150,000	175,000	201,000
Total	1,145,000	1,739,000	2,847,000	3,038,000	3,366,300

alternative (reduction of space for landfilling) and, furthermore plastic materials take from 100 to 400 years to degrade in the environment. Incineration might release, in addition to CO_2, different toxic particles to the atmosphere.

Mechanical recycling seems to be the best solution, but still it needs to have highly homogeneous plastic waste in order to add real value for further applications.

A better view on use and end of life of different categories of plastic waste in some of the "high users" European countries (such as Italy, France and Spain) are reported in Table 3.4 [1].

Table 3.3 Recovery route and end-use (tons) in Western Europe in 2002 (APME, 2004)

Type of recovery	Agriculture	Total
Total available plastic waste collectable	311,000	26,607,000
Landfill and incineration (without energy recovery)	145,000	12,817,000
Energy recovery	1000	4,678,000
Mechanical recycling within Europe	149,000	2,466,000
Mechanical recycling for export	16,000	341,000

Table 3.4 Yearly agricultural plastic consumption in some European countries and waste generation in ktonnes

Country and reference year	Spain Year 2004	Italy Year 2005	France Year 2006	UK Year 2003	Greece Years 2002–2003	Finland Year 2006	Cyprus Year 2007
Total APW[a] (tonnes/year)	**74.5**[c]	**230**	**170**	**76.1**	**28**	**8.5**	**0.7**
Total AP[b] (tonnes/year)	235	383	–	–	–	–	–
Greenhouse films	41	58	7.5	5	3	–	0.2
Low tunnel films	5	31	8	–	2	–	0.03
Mulching films	30	42	1.3	–	4	–	0.01
Direct coverings	49	–	–	–	–	–	–
Nets	–	4	0.6	11	2	–	–
Silage films	16	8.5	13	25	–	2	–
Bale wraps	–	–	7.5	1.5	–	5	0.16
Irrigation pipes	75	85	1.8	–	1.5	–	0.25
Fertilizer sacks		12	10	12	8	–	–
Pesticide cans	–	2.5	6.5	2.5	0.05	1	–

Reprinted from Ref. [1] with permission from Scarascia Mugnozza
[a]APW Agricultural Plastic Waste
[b]AP Agricultural Plastic consumption
[c]The quantity is related to: greenhouse, low tunnel and mulching films

Furthermore, according to the DEGRICOL Projects, agriculture and horticulture sector generates about 1.2 Mt of plastic waste per year in the Europe, where almost 30% (336 kt) are coming from accessories such as nets and clips for plant support and plastics pots and trays for greenhouses and bed seeds. Though these products are indispensable for farmers around the world, they also are adding to the ever-increasing problem of plastics disposal [5].

Despite the European Union in the last decade having put a strong focus on policies to reduce, properly dispose of, and to avoid environmental contamination from waste, still the consequences of the large amount of agricultural plastics used are the dependence on oil production and the production of waste that, if not properly collected, treated, and recycled, pollute the rural areas and release harmful substances in the environment. Nevertheless, sometimes plastic waste is illegally burned, abandoned, or plowed into open fields [1].

According to the waste hierarchy the best practice toward waste is to avoid its generation in the first place [6]. Biodegradable materials, especially for agriculture applications ending their life in the field thus avoiding "white pollution," seem to be one of the better solutions to match this requirement.

3.1.2   The Role of Research in the Development of Innovation in Biodegradable Tools for the Agricultural Sector

Political indications on waste prevention and management, necessity of using plastics to increase productivity of agriculture and increase the level of protection, increase awareness of environmental protection, local presence of innovative industries, availability of new materials, and development of standards. All these aspects have contributed in leading European research toward the development of innovative biodegradable and compostable solutions for agriculture and landscaping to reduce the dispersion of plastic waste in the environment.

In the last 10 years much research and development as well as dissemination projects have been financed. Thanks to these projects, new biodegradable applications have been studied in detail and made available in the market as an alternative low environmental impact for growers (e.g., geotextile mulches, clips, and twines), while other are still at an advance state of industrial development (e.g., irrigation drippers, silage films) [7]. In Table 3.5 some of the EU-financed projects aimed to develop biodegradable and compostable applications for agriculture are reported.

Together with these, other projects at regional level have funded applied research in new solutions for reduced impact for the environment.

Table 3.5 Main EU-financed projects in which biodegradable and compostable tools for agriculture were developed

Type of project	Title	Years	Products	Country
FP5-INCO	COTONBIOMAT [9]	2001–2005	Processing cotton into biodegradable materials for agriculture	France
FP6-2003-SME-1	PICUS [10]	2005–2007	Fibers for 100% biodegradable twines and nets for packaging	Spain
FP7-AME-2008-1	HYDRUS [11]	2009–12	Drip tubes	Spain
FP7-KBBE	FORBIOPLAST [8]	2008–12	For agriculture: pots, base for slow-release fertilizes, Tomato yarns	Italy
FP7-NMP-2007-SME-1	BIOAGROTEX [12]	2008–2010	Agrotextile for agriculture	Belgium
FP7-SME	DEGRICOL [5]	2012–2015	Agriculture accessories Pots, clips, tutors and nets	France
CIP 2007/2013	DRIUS [13]	2013–2015	Drip tubes	Belgium
LIFE03 ENV/IT/000377	BIOCOAGRI [14]	2006–2009	Applications for agriculture; Sprayable film, biodegradable mulch film	Italy

From Cordis [7]

3.2 Case History: The Greenhouse

3.2.1 Plastic Materials in Greenhouses

The greenhouse system is by definition the most developed system where plastic materials are used and combined to increase the quality and quantity of production in a fully controlled way. For greenhouse use, many of the plastic solutions are required to have a long life (e.g., covering films), and some of plastic waste is not too contaminated at the end of its life (e.g., irrigation tubes) and therefore can be easily collected and disposed of appropriately (Table 3.6).

Some other plastic applications (e.g., clips or twines/yarns) are commonly used only for the growing season of crops such as tomatoes or bell peppers (mostly up to 12 months) and then need to be removed, and in many cases they are highly contaminated with plant residues. This makes a proper end of life for both plastic waste and organic residues almost impossible.

Table 3.6 Main plastic
materials in the greenhouse
system

Covering films
Irrigation tubes
Twine
Trellising accessories (clips, support hooks)
Hydroponic reservoirs (bags, containers)

Clips and twines/yarns represent good examples of how the transition to an innovative biodegradable and compostable material can improve the environmental sustainability of this system, with the possibility to produce organic matter and to return in the soil via composting.

Tomato is the second most grown vegetable in the world, and about 12% of tomatoes are produced in Europe. According to the FORBIOPLAST Project the tomato yarn required in Europe (EU-27) is approximately 350 kg/ha and it has been estimated that approximately over 100,000 tons of this string could be substituted with biodegradable and compostable materials in Europe [8].

The materials used for "traditional" yarns can be both plastic, such as polypropylene (PP, the most used plastic material for this application) and polyethylene (PE), as well as natural fibers, such as jute and raffia.

Biodegradable and compostable alternative are mainly produced in polylactic acid (PLA).

The main features required for a good performing yarn are as follows:

- High performance in a warm and humid environment (greenhouse) for a crop season bearing increasing plant weigh, which means that the remaining strength of the yarn has to be enough not to break;
- High resistance to UV radiations;
- No creep. Creep is the permanent deformation of loaded materials during extended period of time. A yarn should not be modified and stretched during the crop growth and lose its firmness. A high material creep may lead, in a greenhouse, to the possibility of losing tension, of blocking the passageway and therefore interfering with the harvesting operations and reducing the light between the plants.

PLA is a very suitable material for this application since, due to its chemical nature, it is highly resistant to UV radiation (or UV-transparent), leading to no reduction in strength during the crop cycle and it does not creep, keeping the plants standing straight during the whole season. Furthermore, PLA has a low moisture absorption and it is characterized by a lower specific gravity, making it lighter in weight.

PLA yarns can be disposed of as green waste and composted within the time required by biodegradable and compostable materials (EN 13432 [15]), which is 180 days in composting conditions (50 °C and humidity). In order to have a full biodegradable and compostable system for greenhouses crops other trellising accessories (such as clips, winding hooks, truss) should be produced using these materials.

In some European countries (e.g., the Netherlands, as reported below) growers can save money and time using biodegradable and compostable materials due to a lower taxation on green waste compared to mixed waste.

3.2.2 Case History: The Business Case of Compostable Clips in Dutch Greenhouses

The Netherlands has 10,000 ha of greenhouses, of which 4830 ha are used for vegetable crops. The main vegetable crops are tomatoes (1780 ha), peppers (1160 ha), and cucumbers (600 ha).

The crops are cultivated in a soil-less system (hydroponics). In most cases mineral wool is used as a growth substrate inside plastic grow bags. There are many growing systems for the plant, but most of them have in common that twines and number of plant support items (trellising accessories) are being used. Plant support items are for example truss supports, hooks, and clips. Although these items are beneficial during the growing season, they have an important drawback at the end of the season, when the plants need to be removed from the greenhouse: the composting of the organic waste from greenhouses is complicated because of the plastic contamination of the green waste.

Waste management companies in the Netherlands have introduced differentiated gate fees depending on the (plastic) contamination of the waste, in order to have from the gate of the composting plant a good source of feeding organic material. Some typical gate fees are given in Table 3.7.

The difference in gate fess can be explained by the fact that non-compostable twine is easier to separate from the organic waste than clips. For the removal of clips the composter has to decrease the diameter of the screen in the sieving step at the end of the composting, because of the small size of the clips. This implies that more compost is sieved out together with the plastic clips, leading to less sales of quality compost and thus reduced revenues and increased costs for processing of the residuals. Therefore, the introduction of compostable clips is the most beneficial from an economic point of view to start with for usage in greenhouses.

The total amount of waste from greenhouses is 30–50 tons per ha, depending on the crop. So the amount that can be saved on the gate fee ranges from 1950 to 3250 euro per hectare. The savings can be (partly) spent on the increased price of compostable rope and clips (or other support items).

Table 3.7 Typical gate fees organic waste from greenhouses in the Netherlands

Type of waste	Gate fees (€/tons)
Clean organic waste	30.00
Organic waste incl. plastic twine	60.00
Organic waste incl. plastic twine and plastic clips	125.00

The number of clips used varies a lot per crop, and growing system. In general it can be said that in the case of tomatoes much more clips (up to 25 per plant) are being used than in the case of pepper and cucumber. When this is the case the additional costs of the compostable clips cannot be justified compared the benefits. In case of the pepper and cucumber (compostable) clips can be used to attach the stems of the plant to the twine. This is faster and easier than binding the twine to the plant. And in this case the additional costs of compostable clips are lower than the extra costs of waste management.

Clips

The primary function of clips is to hold the plants upright, attached to the rope. This implies that

- The clip will not break;
- The clip will not slip on the rope;
- The clip will not damage the plant;
- The clip will withstand the climate conditions in the greenhouse;
- The clip should be easy to attach to the rope/plant.

The clips are produced via injection molding. This means that material should fulfill the technical requirements for injection molding. Important parameters for the processing of materials are the density, the thermal transition temperatures (like the glass transition temperature), the melt flow rate, and the shrinkage. For conventional clips PP is being used. This material is well suited for injection molding and fulfills the functional requirements for clips. However, a big disadvantage is that the PP is not compostable.

The requirements for a product to be called compostable are set in the standard EN13432 [15] and EN14995 [16]. These standards require that the clips should be biodegradable and disintegrate in a composting system (within a certain time frame) and that the compost produced meets the quality criteria. Furthermore, the clips should be recognizable in order to be able to check whether individual products fulfill the requirements.

Trials

In Fig. 3.1 the compostable clips are shown.

In order to find the right properties for well-performing clips, several biodegradable and compostable materials for clips were tested. They were first tested in a laboratory environment. A basic test for the functional properties which simulates both the strength and the slip behavior on the rope under load was performed as a screening method. After these tests were completed the clips were tested in the greenhouse for one season in various crops.

Simultaneously with the trial in the greenhouses composting tests were conducted both in the lab and in practice. The laboratory tests were performed according to the EU standards on compostability EN13432 in a certified laboratory.

The tests to evaluate the end of life in compost were performed by facility specialized in waste management of greenhouse waste in the Netherlands. The first cycle in this facility takes about 6 weeks. The length of the second cycle depends

Fig. 3.1 Compostable clips

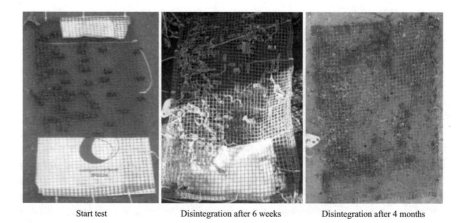

Start test   Disintegration after 6 weeks   Disintegration after 4 months

Fig. 3.2 Composting trials for biodegradable and compostable clips

on the plastic contamination. Without any contamination the cycle time is about 4 months. If the waste is contaminated with plastic ropes (PP) and clips (PP) the cycle time is about 15 months. The compost is sieved at 15 mm after the second cycle, in order to separate the plastic contaminants.

In Fig. 3.2 the composting trials are shown: compostable clips were put into a plastic net and then placed into the composting heap for the period of time required by the EU standard. As a reference material some certified compostable bags were included in the same net as the clips. The test showed that the clips disintegrated (as well as the bags) very well within 4 months. Based on both composting tests a certification was acquired.

The recognition of the clips at this stage of development was the main issue to be tackled before the product could be introduced on the market. Since the product is very small the compostability logo is very difficult to be recognized. Several stakeholders (the grower association, the waste management company, producers, and the bioplastic association in the Netherlands) agreed on a protocol for acceptance of the waste (including the compostable clips) from greenhouses for the lower gate fee. This protocol gave the possibility for the waste management companies to check the compostability of the clips in advance, even during the growing season. In this way a maximum security of absence of contamination in the waste from greenhouses was assured.

The introduction of the compostable clips shows that the compostability can give an advantage not only from an environmental point of view (less contamination in the compost), but also from an economic point of view (less costs for growers and the composting companies). The differentiation in gate fees makes possible internalized external costs and the development of the protocol for acceptance of the compostable clips in the composting process played an essential role in the introduction of compostable clips in the Netherlands.

3.3   Case History: The Orchard

3.3.1   Plastic Materials in Orchards

See Table 3.8.

Modern orchards more and more use plastics to better control and protect the trees. Mulch films can be used when the young trees are planted together with drip irrigation. The orchards are protected with nets against hail or insects, and in many cases young trees are sheltered with plastic nets against possible attack of animals in the first years. Plastic wires might be used in some cultivation systems to direct the young shoots in the first seasons to the right position to create the adequate levels of branches and periodically new shoots may need to be positioned, after pruning the old less productive branches.

From the 1980s lower chemical impact agriculture (specifically Integrated Pest Management, IPM) has introduced a different approach and strategies of pest control, in which synthetic pesticides were reduced, thanks to an optimization of timing of treatments. Pheromones have played an important role in IPM.

The term pheromone was first proposed in 1959 by Karlson and Lusche. Pheromones are volatile substances released in the environment by an insect and received by another insect of the same species, where they induce a reaction. The effect on behavior can be immediate ("releaser" effect) or retarder on the physiological processes connected to reproduction or development ("primer" effect) [17].

Table 3.8 Main plastic applications in the orchard system	Mulch film/geotextile
	Ropes
	Pheromone dispenser
	Drip tubes
	Anti-hail nets
	Shelters

Releaser pheromones can be classified according to the effect they are producing on the receiving individual, such as sexual pheromones, aphrodisiac pheromones, aggregating pheromones, dispersion pheromones, and alarm pheromones [18]. Pheromone traps for monitoring and dispensers for avoiding the mating started to appear in the fields: both devices are mainly made out of plastics.

Some of the plastic materials used in the orchard can be properly collected and disposed of (nets, drip tubes, shelters). Others have a high risk of being abandoned in the fields on the trees and remain there. Among these are the pheromone dispensers.

The case history will be focused on the first example of biodegradable pheromone dispensers used in Italian orchards against some of the main pathogenic insects of apple trees. In this case the end of life will be the field, where the biodegradable plastic will be transformed in CO_2, water, and biomass by the microorganisms in the soil.

3.3.2 Case History: The Business Case of Biodegradable Pheromone Dispensers in Italian Orchards

Mating disruption is a method to control insects based on the use of sex pheromone. The first commercial mating disruption pheromone was introduced in the late 1970s against the pink bollworm of cotton [19]. Since the first commercial use, this technique has grown steadily and has helped growers against a wide variety of insect pests of various agricultural and forest crops [20].

Mating disruption technique consists of dispensing relatively large amounts of synthetic female sex pheromones via pheromone-impregnated dispensers per hectare of crop, so to reduce/suppress the male insect ability to locate the female for mating, using false trails. This technique is particularly suitable for sustainable/organic agriculture because it minimizes the deleterious effects on non-target fauna, including beneficial arthropods, leading to a reduction or elimination of insecticide treatments. In addition, since mating disruption may be combined with low-impact microbial control and it can be applied to control pests in the organically managed agroecosystems.

Considering that in Italy pheromones mating disruption technique could be applied to more than 120,000 ha, and it has become a fundamental pest control method for some crops, such as apple and pear trees, apricots, peach, and vine. For example, in Trentino Alto Adige, one of the most dedicated cultivation Italian areas for apple trees, it has been estimated that 80% of the trees are treated with pheromones.

Generally the pheromone dispensers used are made of plastic material and have different designs (impregnated threads with metal wire, small containers with liquid pheromones, impregnated plastic hooks). The dispensers are placed on the tree branches once or more times in a season, and generally left there. In some cases the dispensers can drop on to the soil or sometimes can be removed while pruning, but it is very frequent to see, in winter, fruit trees with used dispensers still placed on the branches.

The amount of plastic left in Italian orchards by the non-collected plastic pheromones dispensers has been calculated by considering the average weight of each dispenser, the number employed per hectare, a 120,000 ha estimated area, where mating disruption technique is used, and by considering 100% of dispersion in the environment. To make this amount of plastic easy to be visualized in terms of white pollution, it was converted into the equivalent number of plastic bags that will remain in a field each year. In Table 3.9 the details of the calculation are reported [21].

It is clear that even for a small, almost invisible amount of plastic, such a few grams pheromone dispenser can have an important impact and be a source of plastic pollution, if not removed at the end of its use. According to this scenario 100 to over 300 tons of plastic per year are left in the area where the mating disruption technique is used.

In recent years a fully biodegradable plastic dispenser has been developed in Italy in order to eliminate the plastic waste produced in orchards. These dispensers can be used for a type of mating disruption technique defined as "sex disorientation," enhancing the overall sustainability of this method. They have been authorized in organic farming since 2006. In Fig. 3.3 a biodegradable pheromone dispenser in use dispenser on apple tree against *Cydiapomonella* and *C. molesta* is shown.

The efficacy of the use of biodegradable pheromone dispensers in the "sex disorientation technique" has been widely demonstrated by several studies and commercial applications, against some of the most dangerous pest insects of fruit trees (pome and stone fruit) such as *Grapholita* (*Cydia*) *molesta* and *funebrana*, and *Anarsialineatella* [22].

The mating disruption technique using biodegradable pheromone dispensers is characterized by important environmental externalities, such as the reduction of insecticides which impact positively the quality of the production. This technique avoids the creation of resistance in pest insects and increases operator safety, and finally avoids the dispersion of plastic in the environment. For these reasons in Italy this technique has been included in the "Environment measures" of the agriculture European Common Organization Market (CMO) founding scheme for growers.

More recently the mating disruption technique using biodegradable materials has been assessed using a new type of dispenser: a pheromone-impregnated-thread dispenser [23]. The thread has been optimized against *S. littoralis* on a short-cycle crop like spinach. This technique seems to be particularly suitable for processing spinach, because it can minimize the amount of insecticide residues in a crop characterized by a short cycle, improving the ecological impact of the cultivation technique. This biodegradable, low-dosage, slow-release pheromone dispenser has been tested in the glasshouse and open field also on other crops such as cyclamen and herbs, proving to be effective in decreasing *S. littoralis* males trap catches and leaf damage [24].

Table 3.9 Calculation of the quantity of plastic left in 1 ha by plastic pheromones

Quantity of plastic/dispenser (g)[a]	No. of dispenser/ha (min–max)	Weight of plastic left in orchard (g/ha)	Average weight of 1 plastic bag (g)	No. of equivalent plastic bag left in orchard (g/ha)	Weight of plastic in orchard (kg/ha)	Area covered by mating disruption in Italy (ha)	Total weight of plastic left in orchard (min–max) (kg)
2.74	300–1000	822–2740	12.5	66–219	0.8–2.7	120,000	**98,640–328,800**

[a]Obtained by weighing some of the most widely used plastic pheromone dispensers

Fig. 3.3 Biodegradable pheromone dispenser on apple tree against *Cydia pomonella* and *C. molesta*

It can be easily used for a wide range of crops, such as vegetable or flowers that, unlike orchards, where the dispensers are applied directly to the branches of the trees, are usually lacking suitable supports for an adequate number of uniformly distributed dispensers. This pheromone-impregnated-thread dispenser could be a solution for a crop like spinach, cultivated on large areas, minimizing the costs of distribution of the pheromones.

3.4   Case History: Agrotextile Applications in Forestry and Agriculture

Geotextiles and agrotextiles are important and increasingly used applications for different sectors in agriculture and reforestation due to their characteristics, which enable them to be used in different conditions to those of extruded mulch films.

Geotextiles are thin, permeable materials used primarily in civil engineering applications to improve the structural performance of soil and of works, such as road pavements, even if in recent years their use is increasing in the agricultural sector for revegetation and re-construction of forest habitat for landscaping. The term agrotextile is used for woven, nonwoven, and knitted fabrics used for agriculture, horticulture applications, including livestock protection, shading, weed and insect control, and extension of the growing season. The main applications in agriculture of geotextile materials are reported in Table 3.10.

Table 3.10 Examples of the main application of agrotextiles

Sunscreen nets	Packaging sacks and wrappers
Wind shield nets	Fishing nets
Bird nets	Cut grass collection bags
Crop covers	Underlay fabrics
Harvesting nets	Under support nets
Insect nets	**Ground covers**
Anti-hailstone nets	**Mulch mats**

Presently, the agrotextile market is dominated by synthetic materials mainly produced from PP, but also PE and to a lesser extent other petrochemical polymers such polyamide (PA) and polyethylene terephthalate (PET) are used.

Natural fibers are also used such as jute, kenaf, coir, cotton, sisal, and other natural materials such as palm leaf, wood, and spit bamboo have applications as well. Biodegradable polymers used for agrotextiles are mainly polylactic acid (PLA) and to a minor extent polyhydroxybutyrate (PHB), polybutylene succinate (PBS), cellulose esters, soy-based plastic, starch plastic, and biobased resins from functionalised vegetable oils [25].

Geotextiles can be divided generally speaking into two main categories: woven and nonwoven, even if a number of specialty products are appearing on the market.

Nonwoven are mainly used in the market of drainage, linin system, and asphalt overlay fabric. Woven are most frequent used for geotextiles for soil stabilization and separation, sub-grade and base reinforcement, and silt fence.

The production of geotextiles is a relatively simple process. The majority of nonwoven take the form of a heat-bonded continuous filament or needle-punched filament or staple, made of synthetic polymer. These are relatively lightweight fabrics weighing from 120 to 250 g per m^2. Woven fabrics range from inexpensive lightweight woven silt PP film to complex multifilament polyester yarn fabrics weighing to 2000 g per m^2.

Agrotextiles offer very attractive applications and the volumes in this market area are high and fast growing (Table 3.11) [26].

Among the main areas where agrotextiles are used Western Europe plays an important role with over >200 ktons/year [27].

The case history will be focused on *mulch films (groundcovers)* and *mats* for forestry, natural engineering, and landscaping (such as erosion control of slopes along motorways, railways, revegetation areas, green public areas, etc.).

The estimated European market for this application is 12,500 ha, on which 85% is PP agrotextile mulches and 15% organic mats, with important prospects for growth [28].

The main advantages of using a geotextile mulch, instead of extruded films, are linked to the different environment where these materials are placed and their functionality. The geotextile mulches are used primarily to preserve and enhance the growth of trees and bushes planted to recreate a natural habit or create a mitigation area, or in green areas, avoiding the growth of weeds. For these

Table 3.11 Identified and forecast by DRA report of world consumption of agrotextiles in volume in 2005 and 2010

Regions	Volume of consumption ktons	
	2005	2010
North America	150	152
USA	135	137
South America	117	142
Brazil	44	53
Western Europe	250	261
Eastern Europe	72	93
Middle East	51	63
Central Asia	14	16
South Asia	195	269
India	154	215
North East Asia	504	630
China	389	509
South East Asia	158	205
Africa	93	110
Oceania	13	16

applications irrigation is not always a possibility and therefore the materials need to be permeable to rain water, which should efficiently reach the roots and remain in the soil longer than in a natural bare soil situation.

The *advantages* of ground cover for natural engineering or green areas can be summarized in

- Limitation of weeds growth around the young trees/shrubs. Especially in the first 2 years the weeds can strongly compete with the trees;
- Reduction in the maintenance costs (no need of herbicides application, no need of mechanical weeding), especially in the first years;
- Alternative to herbicide treatments, especially in protected areas (natural reservoirs) or such as city parks and therefore reduction in using chemical molecules;
- Positive influence of mulching materials on the evapotranspiration and retention of moisture in the soil. Agrotextile mulches are both permeable to rain water but, at the same time they reduce the water evaporation from the bare soil to the atmosphere. This, together with an increase in the soil temperature, enhances the tree's growth and development and also allows to use younger (cheaper) plants;
- Combining agrotextile mulches and shelters and/or protection nets increases the protection of young trees against animals and therefore reduces the missing plants and the costs for replacing them [28];
- Some agrotextile (mainly woven) can be pre-sown.

However, some *negative aspects* are connected to the usage of plastic geotextile can be found in

- long lifespan in the field and high possibility of leaving plastic residues in the soil; agrotextiles generally need to cover the soil and retrain their properties for 1 up to 5 years. In most cases after this period of time it is practically impossible to retrieve the materials from the fields, or, whenever possible, the materials will be polluted by a vast amount of organic material and sand, making efficient recycling and even combustion with energy recovery extremely costly and not attractive;
- together with agrotextiles often other plastic applications are used (such as shelters), and also in this case they will not be removed from the field, with a negative impact on the developing habitat and visual impact on the landscape.

In the sequence of pictures in Fig. 3.4 there is a clear example of plastic residues remaining in the soil after a revegetation of an area next to a railway line in Italy. The trees were planted in the late 1990s and in 2014 the plastic mulch is still present at the base of the grown trees [29].

The main functional characteristics of a geotextile material in order to be able to guarantee the expected performance in the field should be

- resistance in outdoor conditions for a minimum period of 3–5 year, which means a good UV stability (PP fabric for ground cover is over 36 months with a constant behavior);
- mechanical characteristics (strength—kg/cm) enabling commercial installation. The reference value for a PP fabric for ground cover is 20 kg/cm;
- not too heavy, but able to control the weeds and remain on the soil. A low material weight together with good ground coverage capacity will speed up the installation;
- low flammability;
- For biodegradable materials: non-ecotoxicity effects in the soil during degradation and proved biodegradability and compostability in the final environment (no "white pollution" with fragments remaining for long time).

(a) Beginning of 1990s. Mulch film and young trees   (b) After 2 years   (c) Year 2014. The forest and still the remaining plastic at the base of the trees

Fig. 3.4 Use of plastic geotextile in open field. Reprinted from Ref. [29] with permission from G. Sauli

In many cases agrotextiles can be based on natural fibers (mainly jute), but in general these products are degrading so fast in the natural environment that their lifetime is usually limited to one or maximum 2 years and a relatively higher weight per m^2 (up to 1.0 kg/m^2) is required in order to compensate for the fast degradation [12].

The main biodegradable and compostable biopolymers used for these applications and available on the market are PLA, or PLA/PHB and in some cases combined with natural fibers. PLA has been proved to be a good alternative to plastic PP and PET or PE fibers since it has a number of features that are similar to many other thermoplastic synthetic fibers, such as controlled crimp, smooth surface, and low moisture regain. Its mechanical properties are similar to those of conventional PET, and also a comparison can be appropriate also to PP. Concerning the fibers both filament yarns and spun yarn can be made in the same way as with PET. The tenacity of PLA fibers is higher for natural fibers and they are also relatively unaffected by changes in humidity at ambient temperatures. Finally although PLA is not a 'non-flammable' polymer, the fiber has good self-extinguishing characteristics and it is characterized by a higher LOI (Limiting Oxygen Index) compared to many other fibers, meaning that it is more difficult to ignite and requires a greater oxygen level.

A comparison among performance of synthetic (PP fabric), organic, and biodegradable (PLA fibers) groundcovers available on the market is reported in Table 3.12 [28].

In general, biodegradable ground covers made from biodegradable and compostable materials can meet the main features of a synthetic material; furthermore, intrinsic positive properties of the biobased polymers such as low flammability, lightfastness, or intrinsic preservation properties can boost technological advantages, leading to major economic, and technological benefits in industrial implementation.

A comprehensive study addressed to optimize fully biodegradable and biobased polymers as alternatives to synthetic and natural fibers as well as to improve the commercial offer has been financed under the FT7 scheme.

The EU founded BIOAGROTEX Project (*"Development of a new agrotexile from renewable resources and with a tailored biodegradability"*) aimed to develop

Table 3.12 Main performance characteristics of a synthetic, organic, and biodegradable groundcover

Parameters	Synthetic	Organic	Biodegradable
Weight (g/m^2)	85	1000–1400	110
Thickness (mm)	1	10–20	1
Strength (kg/cm)	20	2	10
Width (cm)	Up to 500	Up to 200	Up to 400
Shelf life (years)	>36 (constant)	12–24	>36 (constant)
End of life	Fragmentation	Biodegradation	Biodegradation

Source Verlinde [28], modified

Table 3.13 Biodegradable agrotextiles available on the EU market, 2015

Product type	Commercial name	Producer	Material	Main characteristics	Lifespan	Main applications
Woven ground cover	Duracover[a]	BONAR Technical Fabric (F)	PLA	Extruded PLA tapes certification biobased Vincotte; conformity to ISO 12952-1:2010	3 years	Covering land besides highways or railways, for weed control in green areas
Nonwoven geotextile	Hortaflex[b]: HortaflexThermo, Horthaflex plus (with seeds), Hotaflex 400 and 300	De Sadeleir	PLA	PLA fibers Weight: 150 g/kg or up to 300–400; UV stability superior to PP and PET; natural flame retardant	3–5 years	Horticulture, geotextile, weed control, soil erosion
Jute-based ground cover	Zelotec[c]	La Zeloise	Mix of natural fibers and biopolymers	Enhanced durability compared to jute; no phytotoxic effects in the soil; good mechanical properties and UV stability		Ground cover, for protection of slopes, viticulture, forestation, green areas, fruit trees
Woven textile Ground cover	Ökolys[d]	Beaulieu International Group	PLA	Fibers: EN13432 compliant	>3 years (EN 14836/ ISO 13934-1 compliant)	To protect plants and soil against erosion
Woven textile ground cover pre-sown felt	Tenax FPV[e]	TENAX	Cellulose	Cellulose fibers 100% biodegradable; contains both seeds and fertilizers; biodegradation from 3–5 months depending on the climatic and land conditions; 250 g/m², 2 mm thick		Sowing grass in green areas, against erosion on slopes; lawn sown[a]

Source Leaflets and web sites of producers mentioned in the table

[a]http://www.bonar.com/media/1562/brochure-duracover-english.pdf

[b]http://www.dstextileplatform.com/functions/content.asp?Pag=27&pnav=4;15;19;&prod=14

[c]http://www.sustaffor.eu/wp-content/uploads/2015/09/Euratex-Succes-story-4-la-zeloise.pdf

[d]http://www.beaulieutechnicaltextiles.com/en/category/c/2/agrotextiles/s/3/weed-control

[e]http://www.tenax.net/construction/turf-reinforcement-biodegradable-non-woven.php

innovative biobased and biodegradable agrotextiles to reduce the impact on the environment, studying, and optimizing formulations as well as realizing demonstrators and pre-commercial products [12]. The results of the project are readily available biodegradable agrotextile materials (some included in the list in Table 3.13) with technical performance equivalent to the PP fabric. The end of life of the products was thoroughly analyzed in order to verify the behavior of innovative products in the soil and their biodegradation speed. The natural fiber-based materials as well as the PLA-based agrotextiles are evaluated for their durability using the "soil burial tests" according to AATCC 30-2004. This is a biodegradability test for textiles buried in a microbiological active soil at high relative humidity content and at a temperature of 29 °C. The activity of the soil is tested via adding a reference cotton fabric that should be degraded completely after 1–2 weeks treatment. PLA is a fully biodegradable polymer but its biodegradation is widely depended upon temperature (above 50 °C) and hydrolysis [33].

The PLA-based ground cover therefore meets the requirement of standards for compostability such as EN 13432 ("*Requirements for packaging recoverable through composting and biodegradation. Test scheme and evaluation criteria for the final acceptance of packaging*") ASTM D6400, ("*Standard Specification for Compostable Plastic*") [34].

At present different biodegradable agrotextile solutions for ground cover are available on the European market and they can represent an alternative to synthetic materials, avoiding white pollution of soils but with technical performance equivalent to the synthetic ones and improved compared to the natural fibers. The main solutions are listed in Table 3.13, where they are divided by category type, raw material, main characteristics, and applications.

3.5 Developments

Among the wide range of applications of conventional plastic in agriculture, some examples of broadly used applications are reported in the following pages. The two applications, still under development, will have an important impact, once commercially available for the reduction of the overall amount of plastic used in the sector, and at the same time represent a further improvement toward a more sustainable agriculture.

3.5.1 New Perspective of Bioplastics for Irrigation Drip Tubes

Water is for sure a limiting factor in many agricultural areas of the world, as well as a limited resource for the world's population. Agriculture uses up over 70% of all the water withdrawn from aquifers, streams, and lakes) [30]. A correct use of irrigation water is a key factor in a modern, more mechanized, and more intensive agriculture, which needs to meet the rising demand for food and fibers. More and more water availability is a fundamental aspect that has to be solved, together with the inefficiency of convention irrigation systems (e.g., surface irrigation) still used in many countries. The net increase of cultivated land over the last 50 years is attributable to a net increase in irrigated cropping. According to FAO reports irrigated area has more than doubled over the period and the number of hectares needed to feed one person has decreased from 0.45 to 0.22 ha per person [30], as reported in Table 3.14.

According to data reported in the Drius Project [14] in Europe, the percentage of irrigated crop area over the cropland is 7.9%. This percentage means that in Europe there are more than 24 million hectares irrigated with different methods. Of this, 25% of total irrigated crop area (540,000 ha) is irrigated using a micro-irrigation system.

In this framework, drip irrigation or trickle irrigation is an efficient method which saves water and fertilizers by applying the required amount of water locally to the roots of plants. The water is given locally either onto the soil surface or directly onto the root zone, through a network of valves, pipes, tubing, and emitters. It is done through narrow tubes that deliver water directly to the base of the plant. From the 1960s this innovative and revolutionary technique became more and more popular, thanks to its excellent performance (increase of yield and quality of the product) and the possibility to irrigate with an efficient usé of water also of crops in dry areas. Drip irrigation is one of the most used forms of micro-irrigation.

The idea to develop a biodegradable drip irrigation system could also be desirable in order to have a fully biodegradable system, when biodegradable mulch films are used. Ideally at the end of the crop cycle, instead of collecting the plastic waste, biodegradable materials would allow rototilling or plowing in the soil, where they will end their life. It has been estimated that in the Europe around 11,000 million meters of PE micro-irrigated pipes (20,000 m/ha) are currently needed; for sure this is a very big market for pipe manufactures. Moreover, the use of micro-irrigation systems is expected to reduce water consumption by around 60%, i.e., 70,000 million cubic meter per year [14].

In the last decade a number of projects and studies addressed the possibility to produce a fully biodegradable system for short-rotation crops (less than 1 year on the soil, such as tomato, maize, and other herbaceous crops), where the plastic waste problem can be completely eliminated and the quantity of non-renewable resources used for a short-time applications (mulch films and drip tubes) will be replaced by biodegradable and compostable materials.

Table 3.14 Net changes in major land use (Mha), 2010 [30]

Cultivated land	Year		Net increase 1961–2009 (%)
	1961	2009	
Cultivated land	1368	1527	12
– Rainfed	1229	1226	−0.2
– Irrigated	139	301	117

For such a system to be successful it has to present characteristics in terms of performance comparable to those of conventional plastics drip tubes (mainly PE), which are withstand the same pressure applied on the tube during the water flows and external loads as well as the exposure to the open field conditions, in order to exhibit a satisfactory mechanical performance, deliver a constant flow across the length and not deteriorate below the limits set during this operation and high thermal resistance of the materials (in the field the working temperature can reach 60 °C). After its use the material will biodegrade in the soil without any negative impact for the environment. Alternatively the drip irrigation tubes may be removed from the field and put in a composting plant or pile at farm level, where they will be transformed into organic matter (compost) to be then introduced again in the soil system [14, 31].

For at least 10 years, many projects have been focused on the implementation and development of biodegradable drip tubes and dripping tubes system for short cultivation periods (less than 1 year). Many researches developed interesting solutions for this application, but still no commercial product is at the moment available on the market.

Different biodegradable materials were tested in these projects with the objective to implement micro-irrigation systems that are 100% compostable. This will allow the system to be composted at the end of the crop season along with the plants and soil. The main application will be crops of small plants such as strawberries and tomatoes that have short cultivation periods (less than 1 year) [14, 31].

At present no commercial application of fully biodegradable micro-irrigation system is available on the market, having the first successful fully biodegradable system available only at a laboratory and pilot scale: tubes obtained by standard pipe extrusion and drippers via injection molding.

3.5.2 New Perspective of Bioplastics for Silage Covers

Approximately 45% of the plastic utilized in agriculture in Europe is destined for silage packaging [3]. A large proportion of the plastic used for silage is dominated by the use of polyethylene (PE) films that can basically be used only for one cycle. The main type of plastic application used in the preserving forage system is reported in Table 3.15.

The most important factors that can influence the preservation of forage during ensiling are the degree of anaerobiosis reached in the filled silo and its maintenance over the entire conservation period [32, 33]. Silage conservation on farm is based on its successful anaerobiosis obtained with plastic films utilized to cover horizontal silos or with stretch films to wrap bales. These two silage techniques require different performances of the plastic films utilized.

An ideal film to cover horizontal silos should have high mechanical properties (puncture resistance, tear resistance) to resist wind, hail, frost, and handling; thickness ranging from 45 to 200 µm; high impermeability to oxygen (full anaerobiosis is necessary); physical strength properties that can be maintained over a long time period (longer than 1 year) in a natural rain- and sun-exposed environment; UV protection (different degrees of protection in relation to the latitude); and costs related to the necessary quality requirements (not the lowest cost).

These films are made mainly in low density polyethylene (LDPE), or recently have been coextruded with polyamides (PA) or ethylene vinyl alcohol copolymer (EVOH) to improve the barrier to oxygen [34].

The stretch films to wrap bales are made in PE in thickness from 20 to 25 μm and recently proposed to be coextruded with barrier polymers to improve oxygen impermeability [35], as reported in Table 3.16.

The characteristics of stretch film used to wrap bales could be established through different certification procedures that suggested minimal oxygen permeability (<9000 cm^3/m^2 in 24 h with 0.1 MPa pressure), high stretching capacity (with a higher elongation at break than 400%), high puncture resistance (>10 N), high UV stability for a period of at least 12 months, and high adhesiveness to keep the layers together (>0.05 N). Therefore, the two film categories have both the common requirement to have high mechanical properties, high barrier against oxygen, and UV protection for periods longer than 12 months.

Table 3.15 Main plastic materials used in the preserving forage system

Silage films
Bale twines
Bale wraps

Table 3.16 Main characteristics of plastic films utilized to wrap bales or to cover horizontal silos for conventional commercial PE and for coextruded PE/EVOH oxygen barrier films (HOB)

Characteristics	Stretch film for wrapped bales		Horizontal silos		
	PE	HOB	PE100	PE200	HOB
Thickness (μm)	25	25	100	200	130
Oxygen permeability at 23 °C 0.1 MPa[a]	7120[b]	19	1780[b]	846	8.8
Oxygen permeability at 50 °C 0.1 MPa[a]	21,360[b]	45	5340[b]	2538	20.8
Puncture resistance to probe penetration (mm)	20.8	16.7	–	–	–
Force at break (N)	6.4	6.2	–	16	21
Elongation at break, MD (%)	534	716	–	601	1113
Elongation at break, TD (%)	1015	942	–	1381	1176
DGL oxygen permeability at 23 °C 0.02 MPa[a]	1425	4	360	169	1.8

Adapted from [34]

HOB high barrier film; *MD* machine direction; *PE* standard polyethylene film; *TD* transverse direction

[a]$cm^3/m^2/24$ h and 65% RH

[b]From the barrier database

All the plastic films utilized to cover silage are intrinsically difficult to recycle, since they are commonly contaminated by soil, sand, silage, and other organic residues [36]. As other agricultural films, characterized by relatively short usage time (12 months) also silage cover and bale films are highly at risk of improper disposal: many are landfilled or burned in the field [36–38]. Biodegradable alternatives can greatly help also in this case.

The polymers useful to develop biodegradable plastic film to cover silages can be derived from renewable biological sources or from petroleum. In the last decades, bioplastics are an increasing alternative to petroleum-based plastics and can come from a wide range of sources, such as starch (maize and potatoes) and oleaginous plants (rapeseed and sunflower) [39].

The first report on the use of biodegradable plastic derived from petroleum to be used to produce stretch films to wrap bales was in the early 2000s [40]. The prototype films made of Ecoflex co-polyester (BASF) offered good mechanical properties as well as a sufficiently low oxygen permeability in order to satisfy the requirements of stretch films used for silage bales. Films which were stabilized by carbon black also fulfill practically all these criteria. These two types of film were used to wrap silage bales which were then stored beneath a roof or outside, either in the field or in storage. On the one hand, the films were attacked from inside the bales, on the other, on the contact surface between the soil and the film. In order to use co-polyester films for silage bales, degradation should be significantly slower [40] concluding that further improvement of the films is needed through adding an adhesive layer limiting degradation or by chemically modifying the film material.

In 2008 a collaboration between the University of Turin (Italy) and Novamont SpA (Novara, Italy) proposed for the first time the development of new biodegradable plastic films from compostable resins based on renewable resources to cover silage [35]. The objective of this first collaboration was to select materials suitable to produce biodegradable films to cover silage in laboratory scale and the evaluation of thickness requirements to achieve satisfactory conservation quality. The first prototypes were made with a starch-based polymer, known as Mater-Bi® (MB; Novamont SpA), which is the first completely biodegradable and compostable bio-polymer ever invented [41]. These first studies produced promising results identifying the best thickness of the film as 120 μm and showed good silage conservation for 2 months, whereas the degradation of the MB film started just before 4 months of silage conservation. This first experiment [35] encouraged the development of a new blend, derived from renewable sources, which led to the production of blown films to cover silage. These films have been improved for their stability to microbial activity over time, and to have an oxygen permeability that is 41.8% lower than commercial PE films of the same thickness. Therefore, subsequent studies worked on stabilizing the 120 μm MB-based films for longer period of conservation and to outdoor conditions [38]. These first promising results led to develop of a project funded by the Regione Piemonte (POR-FESR 07-13-ASSE I.1.1)—Agroalimentare, project "F&F BIOPACK—Feed and food packaging— Biodegradable films for the environmental sustainability of the agro-food chain" for the years 2011–2013.

From the project F&F Biopack, new stabilized Mater-Bi-based films were developed. The first results of these films to cover silage were published by Borreani and Tabacco [34] who utilized them in pilot trials indoors to cover maize silage, with the aims to establish their performances and to obtain guidelines to develop new biodegradable films to be utilized for farm-scale experiments. The main characteristics that separated these new films from the PE films were the water vapor transmission rate, which was tenfold greater in the MB films than in commercial PE films, and the oxygen permeability, which was more than halved when compared with a LDPE film of the same thickness [34]. The promising results contributed to a step forward in the development of new biodegradable film, which could protect silages for at least 5 months. The most important single factor that can influence microbial and nutritional quality of a forage during ensiling is the maintenance of anaerobiosis and is due to the mechanical characteristics and physical properties (i.e., oxygen impermeability) of a plastic film [33]. If the airtight sealing of the silo is not appropriate, air penetrates the silage, and aerobic microorganisms multiply, thus resulting in aerobic deterioration, and the DM losses in the top 0.5 m can exceed 35% [33]. These results have confirmed that the maintenance of a high degree of anaerobiosis during conservation is crucial for silage quality.

Therefore, the new step was to improve the MB blend to enhance film stability over time, and to evaluate biodegradable plastic films under outdoor conditions. The objective of the next step was to have a biodegradable plastic film to cover silage that is stable to microbial and hydrolysis activities for longer periods than 8 months under natural rain- and sun-exposed conditions, coupled with a high impermeability to oxygen.

The newly developed MB biodegradable films were tested outdoors during winter–spring season by Spadaro et al. [42] in comparison to commercial PE film. The tested films were a single 120-μm-thick (4-m width) light green Mater-Bi-based biodegradable plastic film (MB) and a commercial single 200-μm-thick (6-m width) black-on-white PE, UV protected film (PE). Also the oxygen permeability (D 3985-81 standard method, ASTM, 1981) of the MB film was improved in comparison with PE film (500 vs. 1196 cm^3/m^2 per 24 h at 0.10 MPa at 23 °C, 90% relative humidity (RH)). The WVTR values were 17.4 g/m^2 for MB (for 24 h at 38 °C and 90% RH) and 1.05 g m^2 for PE (ASTM F1249-06 standard method, ASTM, 2011). The results showed a stability of the innovative biodegradable film for about 5 months with a good conservation of silage and the improvements for the future are to further improve oxygen impermeability and the stability of the new MB films to obtain silages with longer shelf life after air gains access to the silo during consumption, by delaying the growth of molds and by reducing their detrimental effect on safety and quality of silage [42].

The results presented encouraged the development of biodegradable plastic films to cover silages in outdoor conditions through the improvement of the biodegradable blends to enhance microbiological film stability over time, and also to evaluate biodegradable UV stabilizer to maintain film performances in outdoor conditions both in winter and in summer conditions.

3.6   Conclusions and Perspectives

In order to give an answer using the available data to the state of the art of applications of biodegradable materials in agriculture, observing that they can be efficient with a better end of life in terms of plastic waste and possible loss to the environment, let us re-examine Table 3.1. The applications in bold are already present on the market and potentially can be substituted with innovative biodegradable materials.

There is a relationship between the duration of use (short) and potential of uncontrolled loss to the environment and the possibility for substitution. This represents the correct adaptation of a biodegradable and compostable material in agriculture, and is also the answer that this technology can deliver to the sector's sustainability.

Protected cultivation films:	Nets:	Packaging:
• Greenhouse and tunnel	• Anti-hailstone	• Fertilizer sacks
• Low tunnel	• Anti-bird	• Agrochemical cans
• **Mulching**	• Wind breaking	• Containers
• Nursery film	• Shading	• Tanks for liquid
• Direct covering	• Harvesting	storage
• Covering vineyards and orchards		• Crates
	Piping, irrigation/drainage:	**Other:**
	• Water reservoir	• **Silage films**
	• Channel lining	• Fumigation films
	• **Irrigation tapes and pipes**	• **Balet wine**
	• Drainage pipes	• **Bale wraps**
	• **Micro-irrigation**	• **Nursery pots**
	• **Drippers**	• **Strings and ropes**

There is still much to do to make these applications commercially available for agriculture, as well as for parks and recreation areas, and those situations where it is not possible or economical to collect plastic; we are on the right course.

In this development, legislation that increasingly aims to reduce waste or help eco-design can do a lot. Environmental externalities and the incorrect use of applications are difficult to be taken into account when it is limited to study and compare the market performance and prices of innovative applications with those of conventional plastics.

In the reported case histories it has been shown how a policy aim to reduce the taxation linked to the implementation of virtuous waste management can efficiently help the agriculture operators to choose compostable products (e.g., compostable clips in the Netherlands). The introduction of innovations would benefit in an initial phase to be directed also by political support in order to compensate and underline the environmental externalities, not always taken into account by the market price.

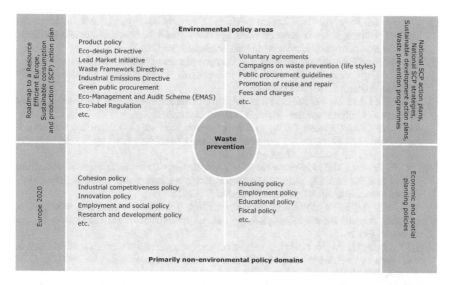

Fig. 3.5 Waste prevention as a cross-cutting policy area. Reprinted from Ref. [6]. © European Environment Agency, 2014

A summary and a possible roadmap for the valorisation of the environmental externalities of the use of biodegradable and compostable applications in agriculture are contained in the EU indications for Waste Prevention in Fig. 3.5 [6]. There the legislative areas favoring the environment are detailed in order to reduce waste production. There is no single way to achieve this, and it is certain that biodegradable materials will have a fundamental role insofar as they fit into the definition of the reduction of waste given by the Waste Framework. Framework: "…'prevention' means measures taken before a substance, material or product has become waste."

Another important factor that can help to highlight the environmental advantages of biodegradable and compostable materials is the possibility that the organic matter produced during composting can be reused in agricultural soils.

References

1. Scarascia-Mugnozza G, Sica C, Russo G (2012) Plastic materials in European agriculture: actual use and perspectives. J Agric Eng 42(3):15–28
2. Jouët JP (2006) Situation of agricultural plastic 2003–2005. In: CIPA Congress 24–25 Oct 2006
3. Vittova K (2013) Latest developments and challenges in the agricultural film market. In: Proceedings on agricultural film 2013. International industry conference on silage, mulch and tunnel films used in agriculture. Applied Market Information Ltd., Madrid, Spain, pp 1–11
4. Plastic Europe (2004) Plastic in Europe. An analysis of plastic consumption and recovery in Europe

5. DEGRICOL Project (2016) Consumer-safe and thermally-stable bioplastic formulation with controlled biodegradation properties for agricultural and horticultural accessories. http://cordis.europa.eu/result/rcn/169614_en.html. Accessed Aug 2016
6. EEA Report (2014) Waste prevention in Europe—the status in 2013, No 9/2014
7. Cordis—Community Research and Development Information Service. http://cordis.europa.eu/home_en.html. Accessed Oct 2015
8. FORBIOPLAST Project (2015) Forest resource sustainability through bio-based-composite development. http://www.forbioplast.eu. Accessed Oct 2015
9. COTONBIOMAT Project (2015) Processing cotton seed into biodegradable materials for agriculture as an alternative to synthetic polymers in Latin America. http://cordis.europa.eu/project/rcn/58742_en.html. Accessed Oct 2015
10. PICUS Project (2015) Development of a 100% biodegradable plastic fiber to manufacture twines to stake creeping plants and nets for packaging agricultural products. http://cordis.europa.eu/result/rcn/46907_en.html. Accessed Oct 2015
11. HYDRUS Project (2015) Development of crosslinked flexible bio-based and biodegradable pipe and drippers for micro-irrigation applications. http://www.aimplast.es/projectos/hydrus. Accessed Oct 2015
12. BIOAGROTEX Project (2015) Development of new agrotextiles from renewable resources and with a tailored biodegradability. http://cordis.europa.eu/project/rcn/89311_en.html. Accessed Oct 2015
13. DRIUS Project (2015) Development of new agrotextiles from renewable resources and with a tailored biodegradability. https://ec.europa.eu/environment/eco-innovation/projects. Accessed Oct 2015
14. BIOCOAGRI Project (2015) Biodegradable coverages for sustainable Agriculture. http://ec.europa.eu/environment/etap. Accessed Oct 2015
15. European Committee for Standardization, CEN, EN 13432 (2002) Packaging—requirements for packaging recoverable through composting and biodegradation—test scheme and evaluation criteria for the final acceptance of packaging
16. European Committee for Standardization, EN14995 (2007) Plastics—evaluation of compostability—test scheme and specifications
17. Benvenuto L, Totis F (2008) Valutazione dell'efficacia argonomica ed economica del metodo "disorientamento sessuale "Ecodian STAR" per il controllo combinato di carpocapsa e tignola orientale del pesco". Notiziario ERSA 4(2008):11–18
18. Schiaparelli A, Raggiori F, Rama F, Ponti GL (2004) Feromoni e trappole – guida per un corretto impiego in frutticoltura e viticoltura, Ed. InformatoreAgrario
19. SucklingD M, Karg G (2000) Pheromones and other semiochemicals. Biological and biotechnical control of insect pests. CRC Press, Boca Raton, USA, pp 63–99
20. Lanzoni A, Bazzocchi GG, Reggiori F, Rama F, Sannino L, Maini S, Burgio G (2012) *Spodopteralittoralis* male capture suppression in processing spinach using two kinds of synthetic sex-pheromone dispensers. Bull Insectol 65(2):311–318
21. Manini M, Accinelli B (2013) Personal communication
22. Accinelli G, Angeli G, Rama F., Reggiori F, Ruggiero P (2004) Impiego del disorientamento sessuale per il controllo di *Cydiapomonella* (L.) in frutteti biologici, Communication at AIIP conference, Cesena
23. Rama F, Reggiori F, Albertini A, (2009) Biodegradable device with slow-release of volatile products having an attractant action for the control of insects. PCT Patent WO2009/090545 A2, p 36
24. Rama F, Reggiori F, Albertini A (2011) Control of *Spodopteralittoralis* (Bsdv.) by biodegradable, low-dosage, slow-release pheromone dispensers. IOBC/WPRS Bull 72:59–66
25. Mohanty AK, Misra M, Drzal LT (2005) Natural fibers, biopolymers, and biocomposites. CRC Press
26. DRA report, World Market forecast (2010) Accessed Oct 2015
27. Euratex (2004) European technology platform for the future of textiles and clothing. A vision for 2020. Accessed Oct 2015

28. Verlinde W (2011) Bioplastics and weed control textiles. In: AMI agricultural film 2011. Oral presentation
29. Sauli G (2015) Relazione sintetica sulle potenzialità di impiego di pacciamanti biodegradabili nei settori dell'ingegneri naturalistica. Personal communication. Feb 2015
30. FAO (2011) The state of the world's land and water resources for food and agriculture—managing systems at risk. The Food and Agriculture Organization of the United Nations and Earthscan. Accessed Oct 2015
31. Briassoulis D, Hiskakis M, Babou E (2007) Biodegradable irrigation system for protected cultivation. In: De Pascale S et al (eds) Proceedings on irrigation systems on Greensys. Acta Hort. 801, ISHS 2008, pp 373–380
32. Woolford MK (1990) The detrimental effect of air on silage. J Appl Bacteriol 68:101–116
33. BorreaniG TabaccoE, Cavallarin L (2007) A new oxygen barrier film reduces aerobic deterioration in farm scale corn silage. J Dairy Sci 90:4701–4706
34. Borreani G, Tabacco E (2014) Improving corn silage quality in the top layer of farm bunker silos through the use of a next-generation barrier film with high impermeability to oxygen. J Dairy Sci 97:2415–2426
35. Borreani G, Revello Chion A, Piano S, Ranghino F, Tabacco E (2010) A preliminary study on new biodegradable films to cover silages. In: Schnyder et al (eds) Grassland in a changing world. Proceedings of the 23rd general meeting of the European Grassland Federation, Kiel, Germany. MeckeDruck und Verlag, Duderstadt, p 202–204
36. Holmes BJ, Springman R (2009) Recycling silo-bags and other agricultural plastic films (A 3875). Cooperative extension of the University of Wisconsin-Extension. http://www.uwex.edu/ces/crops/uwforage/A3875_Recycling_silo_bags_and_other_ag_plastics.pdf. Accessed Feb 2014
37. Bhatti JA(2010) Current state and potential for increasing plastics recycling in the U.S. MS thesis. Columbia University Sponsored by Earth Engineering Center. http://www.seas.columbia.edu/earth/wtert/sofos/bhatti_thesis.pdf. Accessed Sept 2015
38. Borreani G, Tabacco E, Guerrini S, Ponti R (2013) Opportunities in developing novel biodegradable films to cover silages. In: Proceedings on agricultural film 2013. International industry conference on silage, mulch and tunnel films used in agriculture, Applied Market Information Ltd., Madrid, Spain, p 4.1–4.13
39. Momani B (2009) Assessment of the impacts of bioplastics: energy usage, fossil fuel usage, pollution, health effects, effects on the food supply, and economic effects compared to petroleum based plastics. Worcester Polytechnic Institute, Worcester, MA. http://www.wpi.edu/Pubs/E-project/Available/E-project-031609-205515/unrestricted/bioplastics.pdf. Accessed 17 Feb 2014
40. Keller A (2000) Biodegradable stretch films for silage bales: basically possible. Agrarforschung Schweiz 7(4):164–169
41. Bastioli C (1998) Properties and application of Mater-Bi starch-based materials. Polym Degrad Stab 59:263–272
42. Spadaro D, Bustos-Lopez M, del Pilar M, Gullino ML, Piano S, Tabacco E, Borreani G (2015) Evolution of fungal populations in corn silage conserved under polyethylene or biodegradable films. J Appl Microbiol 119:510–520

Chapter 4
Agronomic Effects of Biodegradable Films on Crop and Field Environment

Lluís Martín-Closas, Joan Costa and Ana M. Pelacho

Abstract This chapter describes the state of the art of the agronomic effects of degradable bioplastics used as agricultural films. Current use of bioplastics and certified commercial biodegradable materials, both as granulates and as final products, are introduced. Following, agronomic effects on crops are reported and compared to the routinely used oil-based nondegradable plastics, basically the polyethylene films. Biodegradable films for agriculture were initially developed mostly for mulching application, which still remains the most significant one. Since last reviews published in 2011, new progress and perspectives have mainly arisen regarding the agronomic effects of biodegradable mulching on vegetable crops, not only as films but also as nonwoven biobased mulches. The film mechanical laying and the effects on yield, earliness, product quality, weed control efficacy, microclimatic improvement and film soil coverage and degradation are presented in detail for tomato crops and for other crops where mulching is a common technique (pepper, melon and other cucurbits, strawberry, lettuce,...). Some information is provided for crops not so frequently mulched (broccoli, sweet potato, sweet corn). New findings published on the use of biodegradable films for solarisation are also reviewed, while no significant progress on the use of films for low tunnel covers has been made. Recent proposals for vineyards and future potential application of bioplastics for orchard crops are also addressed. Finally, pros and cons for the adoption of biodegradable films for cultivating crops are discussed.

Keywords Vegetable crops · Fruit crops · Mulching · Nonwoven · Solarisation · Yield · Quality · Biodegradation · Microclimate · Weeds

L. Martín-Closas (✉) · J. Costa · A.M. Pelacho
Department of Horticulture, Botany and Gardening, ETSEA University of Lleida,
Av. Alcalde Rovira Roure 191, 25198 Lleida, Spain
e-mail: martin@hbj.udl.cat

© Springer-Verlag GmbH Germany 2017
M. Malinconico (ed.), *Soil Degradable Bioplastics for a Sustainable
Modern Agriculture*, Green Chemistry and Sustainable Technology,
DOI 10.1007/978-3-662-54130-2_4

4.1   Introduction

The impact of biodegradable films on agricultural crop production is widely recognized. The two main advantages are the adjustment of the films' lifetime to that
of the agricultural life cycle and their suitability to the prevailing recycling systems
in agriculture: in-soil biodegradation and on-farm composting [1]. The use of
biodegradable materials avoids the huge drawback of generating nondegradable
plastic residues that have to be removed from the field and transported to the landfill
or to a waste management system. In certain applications, the disposal of conventional films is not a possibility because, due to contamination with agricultural
debris (soil, plant parts and agrochemicals) and to the UV light degradation [2],
materials are often of low quality, mostly non-recyclable. Cleaning the used film is
usually not economically feasible. In addition, agricultural plastics are dispersed
throughout the rural landscape and are costly and inefficient to collect. In case the
plastic film waste arrives at a collecting point and is accepted by a waste system, it
is commonly delivered to an incineration plant. Unfortunately, incineration plants
able to guarantee no harmful gas emissions to the atmosphere are scarce or even
non-existent in certain locations; in any case they are far away from the farms.
Under the most favourable assumption, the plastic waste will reach an incineration
plant with energy recovery. In such a case, the cost for off-farm recycling is higher
than the cost for on-farm disposal; as a result, the farmers may frequently pile or
bury the waste in an unproductive area or even burn it. The disposal and recycling
of used plastic films for mulching has been extensively discussed recently by
Steinmetz et al. [3].

Innovations in the 80s and early 90s lead to the introduction of biodegradable
polymers and plastics [4]. They provided new alternatives, thus curtailing the difficulties of conventional plastic film waste disposal: waste reduction by the farmers,
making recycling viable and economical for farmers, finding cost-effective ways to
collect, clean and store the material, finding markets for recycled products, etc. [5].
The new materials may have the additional advantage of being totally or partially
biobased. Also, some of the materials are manufactured with agricultural feed-
stocks, thus envisaging new production opportunities for agriculture [6].

The agronomical exploration of these biodegradable films started in the late 90s,
with the first field trials being reported by Weber [7], Nagata et al. [8], and Manera
et al. [9] in Germany, Japan, and Italy, respectively. At the same time, some
research stations in France, Spain and other European countries established their
first biodegradable mulch field trials. In an early publication, Groot et al. [10]
disclosed the raw materials available for manufacturing biodegradable products,
together with its potential for horticultural applications. From these initials to the
present time, information on the effects of biodegradable films on crops and on the
agricultural environment has been widely produced. However, reviews summarizing all these findings are scarce. Kyrikou and Briassoulis [11] reviewed the
biodegradation of agricultural plastic films. The evolution of biodegradable raw
materials available for film converting and the corresponding commercial films for

their applications were described by Martin-Closas and Pelacho [6]. Later on, Hayes et al. [12] reported on the available polymers and polymer blends for biodegradable agricultural mulches.

Updated overviews of the bioplastic raw materials for film converting and the corresponding manufactured products are available in the inventories of the main biodegradation certification bodies. Within the biodegradable registered products for agriculture of the Japan BioPlastics' Association [13], the agricultural end products, the bioplastic materials used for manufacturing them, and the product share are shown in Table 4.1. From the 245 products registered in this inventory, about 75 (31%) are finished products for agricultural/horticultural applications, mostly films for protected cultivation, and occasionally for forestry. Films that may be used for fresh fruit and vegetable packaging or wrapping, but which are not exclusive for the agricultural sector, account for 4% of the products.

The products included in Table 4.1 are certified for biodegradability in composting conditions and, apart from PLA and thermoplastic starch (TPS), the materials are mainly not biobased. Some of these bioplastics (e.g., PBS) can be obtained from biobased feedstocks [14], but this is not the general rule. In another certification body, like the private Vinçotte in Belgium [15], only 17 final finished products under the "Garden, agricultural and horticultural products" entry, and 32 under the "Food packaging flexible" entry are certified under the label "Ok Compost". Among the first entry, mulching films are the most frequent (41%), then pots (29%), landscaping woven and nonwoven ground covers (18%), and finally clips and other horticultural complements (12%). Although Vinçotte does not specify it, it is known that the main materials used for both entries are blends of PBAT with TPS of different origins, with PLA or with cereal flour. All products are partially biobased, and exceptionally some may be made solely from PLA. Similar certified products can be found in the USA Biodegradable Product Institute [16] and in DinCertco in Germany [17]. However, although companies sometimes

Table 4.1 End-bioproducts for agricultural and forest applications registered in GreenPla (Japan BioPlastics Association)[a]

Product type	Bioplastic material	Share (%)
Mulching films	PBAT, PBSA, PBS, BS-LA copolymer, PBLDA, TP Starch, PCL, PLA	51
Films for fumigation	PBAT	3
Sheet/forestry films	PBAT, PCL	12
Bands, tapes, ties	PBAT, PBSA, PBS, PLA	12
Floating covers	PLA	1
Nets	PBA, PLA	4
Yarns, ropes	PBS, PBA, PLA	3
Pots/planters	PBSA, PBS, PBLDA	7
Other products (clips…)	PBSA, PBS, BS-LA, PLA	10

[a]Original produced by authors from source [13]

recommend them as biodegradable for agricultural applications, it has to be emphasized that the certification of the materials or products as compostable does not ensure their biodegradability in the soil. This is the case for products made with Bioflex (BFx) or Bioplast (BP), among others.

The unavailability of a widely accepted international norm with an established criterion for biodegradability is very likely limiting companies to certify their products for in-soil biodegradability. Vinçotte provides the companies with a procedure to test the in-soil biodegradability of the materials and offers the private label "Biodegradable in-soil". Table 4.2 shows the in-soil biodegradable granulates and finished products (films) certified by Vinçotte.

The agronomic performance of biodegradable films on crop and field environment published has been hardly reviewed. Heller et al. [18] provided a practical report on the experimentation on biodegradable mulching performed in Germany from the late 90s to 2007. Effects and specific uses of biodegradable mulches, crops tested and practical recommendations are reported. Martin-Closas and Pelacho [6] reviewed the effects of biodegradable films for different agricultural applications (mulching, solarisation, low tunnels, nonwovens, fruit protecting bags, etc.). Finally, based on some selected papers, Kasirajan and Ngouajio [19] described the agronomic effects of biodegradable mulching films.

The present chapter deals with the agronomic effects of biodegradable films on crop and field environment. It summarizes and broadens the previously reported information, updating it up from around 2010 to the present. Readily in-soil and in-farm composting biodegradable films are included, but oxo-degradable films have not been considered. Past and recent research supports that oxo-degradable materials are not biodegradable [20, 21], nor compatible with the lifespan of most agricultural applications [22, 23]. Films manufactured with blends of biodegradable and not biodegradable films are also not considered as they are not biodegradable. Paper

Table 4.2 **a** Granulates and **b** finished products under the entry of "Garden, horticultural and agricultural products" certified as "Biodegradable in-soil"[a]

Product	Raw material	Company
(a) *Granulates*		
Biolice®/Biofilm® (BF)	PBAT—cereal flour	Limagrain Céréales Ingrédients
DaniMer®	PHA	DaniMer Scientific
Ecovio® (EcV)	PBAT-PLA	Basf Se
Mater-Bi® (MB)	PBAT-TPS-Veg.oils	Novamont S.p.a.
Meredian®	PHA	Meredian Inc.
Mirel® (Mi)	PHA	Metabolix Inc.
So Green®	P (3,4 HB)	Tianjin Greenbio Material Co. Ltd.
(b) *Finished garden, horticultural and agricultural products as films*		
BioBag Agri Film	Mater-Bi®	BioBag International AS
Biofilm Sylva	Mater-Bi®	GroenCreatie
Mater-Bi EF 04P	Mater-Bi®	Novamont S.p.a.

[a]Original produced by authors from source [15]

mulch performance has been recently reviewed by Haapala et al. [24]. Agronomic effects of sprayable or liquid mulches and the applications of biodegradable materials for silage film are developed in another chapter of this book.

4.2 Biodegradable Mulching

Mulching is a multifactorial technique mainly used for the production of vegetable crops that influences many factors of the plant–soil environment. Laying a plastic film on a soil, after seeding or before transplanting, primarily changes the energy and mass balance of the crop system. Comprehensive overviews on the effects of mulching on the microenvironment and on crops can be found in Liakatas et al. [25] and in Tarara [26]. Depending on the properties of the mulch materials (spectral–radiometric, water vapour and CO_2 permeability; mechanical, and biodegradability), and on the interaction of the mulch with sun radiation, soil and crop, their effects vary greatly. Changing the material used for mulching, usually polyethylene (PE), to a biodegradable material can also modify the effects.

The main environmental impact of mulching is on the microclimate of the soil–crop–air environment. It modifies air and soil temperatures, solar radiation reaching soil surfaces, evaporation rate of water from the soil, and gas exchange between soil and air. All these changes in the crop environment have agronomic consequences. The rise of soil temperature hastens the crop development, while soil cooling has the reverse result; an increase of temperature in the air surrounding the crop intensifies growth of the aerial part, increases leaf area and production. Likewise, limiting the light reaching the soil prevents weeds to grow and decreases their ability to compete with the crop for solar radiation. Stabilizing the soil water content prevents hydric stress in plants and provides a more convenient environment for water and nutrient intake by roots. Soil respiration below the mulch releases CO_2, which is likely directed to the mulch openings where the crop plants are; this may result in stimulating photosynthesis.

Mulch effects depend on the environment and on the specific crop. In locations where direct solar radiation is predominant, e.g. southern Europe, the effect of the mulch is to be considerably greater than where diffuse radiation is prevailing, e.g. central Europe. The development of a mulched summer season crop when temperatures are not warm enough may enhance crop development; instead, the effect of the mulch could be detrimental when used for the same crop growing under warmer temperatures. On the other hand, changes resulting from mulching may last all the crop cycle for crops that develop slowly and poorly cover the mulch, e.g. onion, while only early effects are to be expected when crops develop fast to totally cover the mulch, e.g. pumpkin. Additionally, the diversity of cultural practices, such as soil preparation, mulch laying, plant density, crop season, irrigation system, etc., also contribute to the mulch performance with the corresponding crop effects. In summary, depending on these interactions between crop, environment and cultural practices, different agronomic responses are to take place.

Mulch technology has been developed with materials that are converted to films, which are laid on the soil along rows partially covering the field (35–70%), or exceptionally all the area. Traditional mulches are usually thin, 15–40 μm thick films, and are distributed in rolls 0.8–1.8 m wide depending on the farmers' need. Biodegradable films are typically thinner (12–20 μm). In addition to films, low-cost biodegradable nonwovens have been also suggested for mulching to avoid the drawbacks of the current biodegradable textile and film mulch production [27, 28]. The development and evaluation of nonwovens for biodegradable mulching has been faced recently in the USA [12] and also in Poland [29].

Mulching is considered one of the most sustainable techniques in horticulture. The main ground for investing in biodegradable mulching is the need to find an alternative to conventional mulching, based on non-renewable materials and which generates detrimental troublesome residues. Identifying options to the severe disposal constraints of conventional mulching films is not a recent issue. Otey et al. [30] pointed out this necessity, and Swanson et al. [31] characterised the ideal mulching film as one that would degrade at the end of the crop season, and that would completely degrade in a short time when buried into the soil. Some reviews [6, 19] provide a general insight into the agronomic effects of biodegradable mulching.

Crop growth, earliness, weed control, production and quality are the main aspects targeted when mulches are used. The degradation dynamics of the biodegradable mulches are determinant. As previously reported [6], most available information involves MB. Mater-Bi is a family of materials that have been improved for over 15 years and that have been manufactured with different grades: NF 01U/P, NF 803/P, CF 04/P, and EF 04/P. Results from mulches made of other materials: Bioflex (BFx; grade F1130), Biolice/Biofilm (BL or BF), Mirel (MI), Bioplast (BP), Ecovio (EcV; grade M2351) and Ecoflex (EFx) and Ecofilm (EF), are slowly but increasingly available. Experimental nonwoven fabrics for mulching are prepared with PLA and PLA/PHA (75–90 g m^{-2}) blends in the USA [12] and with PBSA (Bionolle; 50 g m^{-2}) [28, 32] and PLA (50–60 g m^{-2}) in Poland [33, 34].

4.2.1 Mechanical Laying and Mechanical Performance

The mechanical laying operation is a key issue for mulching. The film is to be distributed throughout large field areas while simultaneously perforated for the transplants and locating the irrigation tape. Depending on the mulcher, the transplanting process can be carried out in the same operation. By this means the operation costs are decreased and the compacting stress applied to the soil is minimized. During mechanical laying, the film suffers mechanical stresses: tensile stress in longitudinal and in transverse directions, and, depending on how the soil has been prepared and on the way the perforation of the mulch for the transplanting is performed, puncture and tear stresses may occur. During the first stages of the mulched crop cycle, the maintenance of the mechanical properties is significant for withstanding the effects of wind, the puncture or pressure of the soil surface and of certain weeds, and the impact of hailstones. At the time the crop has covered the mulch, the mulch mechanical properties are no longer essential.

Although when purchasing a biodegradable film the easiness of mechanical laying is one of the first concerns for the farmers, mulch field laying evaluation has been poorly addressed. Impediments for laying limit the farmer' acceptance of biodegradable mulches [35]. Nonetheless, many studies evaluating the degradation of the mulch film along the crop cycle lay the film manually, and some others do not mention how the film is installed. However, the initial mechanical stress associated with the laying process may influence the degradation pattern of the material and, to certain extent, the crop performance. It has been sometimes noticed that manually installed films prevailed longer without breakings than mechanically installed films. On the other hand, knowledge on mechanical properties and on the performance of biodegradable mulches has been mostly produced with black films, and very seldom with clear or white films.

To evaluate the potential of biodegradable films for mulch laying, their mechanical properties, mainly tension and tear, are compared in the laboratory with those of PE mulch [36, 37]. It is assumed that if the mechanical properties of the biodegradable mulch are close to those of PE films, the biodegradable film is to perform well when being laid. However, the requirements on the mechanical properties set by the standards are specific for PE-based conventional films. Despite the mechanical properties of the films defined by the standards are unrelated to their installation or use, the success of the installation operation is generally ensured when they comply with these standards [38]. Overdesign of conventional films based on standards may explain this success; when being laid in the field, most PE mulches result to be much more resistant than required. Hence farmers do not need to be concerned about adjusting the mulch tension when laying PE mulch.

The studies mentioned above verified suitable mechanical properties for the biodegradable materials tested: either unaged films made of blends of starch or thermoplastic starch with polyvinyl alcohol [36], or poly ε-caprolactone [37]. Briassoulis [39] used MB films (grade NF803/P, 15–20 μm thick) under field conditions in a watermelon crop (medium Spring–Summer) in Greece; the mechanical performance of these mulches during their useful lifetime was as good as that of conventional thicker PE films (25 μm) in terms of tensile strength in the longitudinal direction. Nevertheless, in the transverse direction, they exhibited very low elongation at break. In a similar field study in South Italy, in a strawberry crop (Fall–early Winter; 124 days), with thicker biodegradable (25–45 μm) and PE (50 μm) films, Scarascia-Mugnozza et al. [40] found less strength and elongation at break in the longitudinal direction for the biodegradable than for PE films; however, mechanical properties were good enough for fulfilling the requirements during the whole crop cycle.

Basically, equivalent results had been previously reported by Martin-Closas and Pelacho [41] with MB films of several grades and thicknesses (NF01U/P: 15 and 18 μm; NF803/P: 12 and 15 μm) used in North–East Spain during a processing tomato crop (late spring-end summer; 110 days). Strength at break in biodegradable MB was slightly less than in PE unaged mulches, while elongation at break was significantly less. The decay of elongation at break at the end of the crop cycle was 32, 53 and 91% for 15 μm PE, MB NF803/P and MB NF01U/P films, respectively. Later on, during the crop, the time course of the mechanical properties was found to be alike for MB, Bioflex and Biofilm [42]. In North USA, in a fresh tomato trial, initial strength and elongation at break was also lower for Ecoflex (35 μm) than for PE films (25 μm) [43]; and, as reported for other biodegradable materials, they lost most of their mechanical properties after 15–30 days in the field. However, the loss of mechanical properties does not usually compromise the soil coverage by the mulch.

Although laboratory evaluations provide a good insight into the potential of biodegradable films for mechanical laying, direct field evaluation is essential. According to several authors [18, 40, 44–46], only little adjustment of the mulcher is required for mechanically installing biodegradable films with conventional mulch laying equipment. Finely prepared soil is required to assess the mechanical installation of the films in the field; hard and sharp items, such as remains from previous crops or big stones, are to be avoided. As compared to PE mulches, for installing biodegradable mulches a reduction of the tension coil is important. In addition, as a general rule, biodegradable thin films should not be microperforated; thus, dismantling of the microperforation cylinder is recommended. Mulch layers without rollers to press the film to the soil are better suited to biodegradable film laying. Small soil particles usually stick to the roller surface and create small holes in the mulch; subsequently, weeds may grow through. For very thin films (e.g. 12 μm) decreasing the mulching speed may be necessary. Also, adequate soil humidity prevents the mulcher wheels pressing the film too hard. As mechanical laying is a critical step for the farmers to adopt biodegradable films, users' guidelines are available to facilitate the initial training [47, 48].

Regarding nonwoven fabric mulching, although Wortman et al. [49] mention that they can be applied with conventional mulch layers, all available reports have used manual laying; to our knowledge, no mechanically laying field evaluation has been published. In contrast with mulch films, nonwoven fabrics are much thicker and heavier, thus limitations may be expected for mechanical laying.

4.2.2 Effects on Vegetable Crops

4.2.2.1 Effects on Tomato

Biodegradable black mulches have been mostly tested with tomato crops, usually by comparing the crop performance with these mulches with that when PE mulch is used. Two different sorts of tomato crops have been tested: processing tomato and tomato for fresh market. Processing tomato is an open field crop, while fresh market tomato can be cultivated in the open field or in big tunnels or greenhouses. Tomato is a summer crop, thus in the northern hemisphere fresh market cultivars are usually grown in the open field from early spring to early fall, while they may be cultivated all along the year in tunnels and greenhouses.

First tomato crop field biodegradable mulch tests were performed with black MB films, comparing different thicknesses (12, 15 and 18 µm) of grades NF01U/P and NF803/P with 15–50 µm LLDPE or LDPE (40 and 50 µm in Italy; 15–25 µm in Spain; 25 µm in USA and Australia), with paper mulches, and sometimes adding a control without mulch. Following, other biodegradable materials appeared in the mulch market, first Biofilm, then Bioflex and later on at the end of the 2000s, Bioplast, Ecovio (grade M2351) and new MB grades (CF04/P and EF04/P). Nonwoven mulches are the most recent alternatives, which are still under testing.

A limited number of publications evaluating *vegetative crop growth* with biodegradable mulching are available. Under Mediterranean continental climate, vegetative development of tomato processing crops mulched with black MB films of the early grades was equivalent to that when PE mulches were used, and higher than that with paper mulches [41, 44, 50]. Usually, the thicker the film is, the better the crop grows. No differences in crop growth between MB grades were found.

Production and yield of tomato cultivated with biodegradable mulching are widely documented. Biodegradable and PE mulches equally enhance crop development; as a consequence, production and yield also increase similarly. Mater-Bi mulch films for processing and fresh market tomato produced and yielded as much as PE in Spain [44, 50–56], in Italy [57, 58] and in the USA [59, 60]. In Spain, Bioflex [42] and Biofilm [42, 55, 56] also yielded similarly to PE and MB. On a fresh open field tomato, Ngouajio et al. [35] reported equivalent results to PE (25 µm) mulches for black Ecoflex (25 and 35 µm), but lower yield when the film was white.

Biodegradable plastic materials have also been compared with paper mulches. Main commercial papers for mulching evaluated in tomato crop have been MimGreen and WeedGuardPlus. Other experimental papers have been occasionally tested (Saikaft, Smurfit Karpan-Liner; Smurfit MG, etc.). No significant differences in production and yield have been found for biodegradable plastics and paper mulches [44, 50, 54, 56, 59–61].

Nonwoven mulch has been tested for fresh market tomato in three different climatic locations in USA [59, 60], both in the open field and under plastic tunnel. Three spunbond nonwoven fabrics (SB-PLA10 white, 657 µm; SB-PLA11 black, 580 µm; SB-PLA12 with PHA black, 390 µm) were compared with PE, MB and paper mulches. In the cooler location, production and yield were higher with MB (black, 15 µm) than with SB-PLA10 white fabric, while equivalent yields were found when SB-PLA10, PE or paper mulches were used. However, in the warmer locations the differences in yield between SB-PLA10 or paper mulches with PE and MB decreased and they even tended to have the highest yield [59]. None of the three nonwoven fabrics resulted in production as much as MB or PE mulches; however, production was slightly lower when using paper mulch [60]. In the open field, Wortman et al. [49] did not identify significant differences in yield between spunbond PLA nonwoven fabrics of different thickness (3–6 mm), mass (71–108 g m^2) or colour and commercial biodegradable MB (15 µm) and Ecofilm (PBS; 19 µm) black mulches.

Earliness is a key agronomic benefit of mulch technology. It is especially relevant for fresh market tomato: it allows farmers to commercialize tomatoes very soon in the season, with better prices for their commodities. Earliness is normally sought by increasing soil temperature with clear mulches. However, for clear biodegradable films there is still a need of improving formulations with UV stabilizers and mechanical properties against weed emergence, thus limited information is available. Candido et al. [58] analysed the performance of a fresh market tomato crop with two MB (black, 12 and 15 µm), a MB "lactescent" (clear diffuse, 40 µm) and a PE (black, 50 µm) film. As compared to the other films, clear MB film increased soil temperature, yield and early production. The black MB films did not affect earliness. However, in contrast, Martin-Closas et al. [44] obtained a significant increase in the number of fruits set per plant with black MB and PE (both 15 µm) as compared to bare soil and paper mulch. When using MB or PE mulches, improved microclimate and water availability were associated to earliness. Moreno et al. [53] did not find significant differences in cumulative yield along time for PE and MB mulched fresh tomato crops, but earliness was not determined.

Quality of tomato production obtained with BD films is well documented for black MB and BF films, and recently somehow also for nonwoven fabrics (SP-PLA11 and SP-PLA12). Main quality parameters for tomato fruits are soluble solids, shape, colour, pH or titratable acidity, firmness, consistency, juice content or dry weight, mean fruit weight and, less frequently, lycopene and total phenolics. Most authors agree that black biodegradable films and nonwoven mulches do not modify the quality of tomatoes produced with PE mulch, which are altogether frequently equivalent to those produced in bare soil plants [44, 50, 51, 56, 57, 60, 62]. As compared with a traditional nonwoven polypropylene (PP) fabric (50 g m^{-2}; 8% PAR transmittance), tomatoes cultured with a nonwoven PLA fabric (61 g m^{-2}; 22% PAR transmittance) had equivalent ascorbic acid and soluble sugars content but lower dry matter percentage and nitrate content [34].

Weed control is one of the most valuable agronomic achievements of mulch technology. Mulches allow controlling weeds with a unique operation from the initial stage of the cultivation, avoiding the use of herbicides or the reiterative mechanical or hand weeding. One round of hand weeding may be exceptionally required if the perforation for transplanting is large. By using mulches, the impact of weeding operations on crop and soil is also minimized. The main disadvantage of mulches is the need of removing them after the crop season; additionally, very specific weeds may not be controlled. Mulching for weed control is specially appreciated in organic farming, where weeds are the most important constraint. In Europe and Canada, the use of biodegradable plastic or paper mulch is allowed, and also conventional plastic mulch, provided that the film is removed away after use. In USA organic systems only paper is currently allowed [59].

Weed control in tomato crop with black biodegradable plastic mulching has been intensively studied. Martin-Closas et al. [50] considered the interaction of the mulch and the crop (processing tomato) in controlling weed growth through the interception of solar radiation transmission to the soil. The opacity of the mulch was essential during the first stages, when the crop had low herbicide capacity because of the very limited development, but at the time the MB (NF01U/P; 18 μm) mulch started degradation, evidenced by the film breakdown, the crop was well developed, interfered with the light transmission to the soil and therefore contributed to prevent weed growth. At 60 days after transplanting the crop allowed less than 20% of PAR sunlight to reach the soil, and under the combined effect of crop and MB mulch less than 8% light radiation was transmitted. For paper and PE, the combination with the crop allowed fewer than 4 and 2% light transmittance, respectively.

In consequence, biodegradable MB mulches which remain without breakings the first 2 months of the crop cycle control weeds. In a fresh tomato greenhouse crop, black MB (NF803/P; 12 and 15 μm) and PE (50 μm) mulches reduced weed density by 50–70%, without statistical differences between them [57]. The emergence of 10 different species was reported, but only *Portulaca oleracea* and *Digitaria sanginalis* grew uncontrolled by MB and PE films. The "lactescent" (clear diffuse) MB film (40 μm) did not control weeds [58]. In a Spanish open field processing tomato, MB (NF803/P; 15 μm), PE and paper mulches allowed very low field PAR transmittance (<2%) and exerted an effective weed control [44]. PAR transmittance of unaged MB, BF and BFx films in the laboratory is under 1% [42]. Thus, it is reasonable to speculate BF and BFx to provide weed control similar to MB, provided that they do not break too early in the field.

Ngouajio et al. [35] considered that for fresh tomato crop performance weed control should be over 85%. They reported very good weed control for two black Ecoflex films (94–97%). However, white Ecoflex mulch allowed a high light transmission and was poorly effective in controlling weeds (35%).

In a processing tomato, herbicide application, manual weeding, paper mulch (brown Saikraft, 200 g m^{-2}), MB (NF803/P, 15 μm), and rice straw mulch controlled weeds as much as PE (15 μm), but technical improvements were required for applying paper and straw mulches [54]. Mulches controlled predominant weed species (*Chenopodium album*, *P. oleracea*, *Digitaria sanguinalis*) except purple nutsedge (*Cyperus rotundus*) which, as previously reported [63], was only controlled by paper mulch. In three semiarid locations with *Convulvulus arvensis*, *C. album* and *Amaranthus* sp. as predominant weeds, PE, MB and Saikraft paper mulches equally controlled weeds [55]; Biofilm mulch (black; 17 μm) and MimGreen paper (black, 85 g m^2) were similarly effective. Other paper (brown Smurfit Karpan-Liner, 120 g m^{-2}; black Smurfit MG, 50 g m^{-2}) mulches were also effective; they performed significantly better than PE and MB mulches for controlling even *C. rotundus* [61]. The improved performance of paper mulches was mainly attributed to high resistance to slow perforation (5–13 times over that of MB and 3–9 times over PE). *C. rotundus* density and soil coverage was higher with MB than with PE mulches. As both of them were equally opaque to PAR radiation, the difference was associated to the higher resistance of PE to slow perforation (ca. 50%).

Different physic–mechanical weed control methods have been compared with the use of PE, MB and paper mulch [64]. Brush hoe was as efficient as mulches in reducing weed biomass per hectare without reducing crop yield. Similarly, in a tomato crop for fresh market under non-irrigated conditions, the combination of mechanical and thermal means (with stale seedbed technique and post transplanting cultivation) with straw mulch may be a real alternative to MB (black; 15 μm) [65].

On the other hand, weed control with a PLA nonwoven mulch (white clear; 640 μm) was far from the efficiency of PE (black; 30 μm), two MB films (black; 20 μm) or WeedGuardPlus paper (230 μm) mulches [59]. However, when carbon black was added to PLA or to PLA/PHA nonwovens [60], light transmission was limited and weeds were controlled as efficiently as with the other mulches.

Mulch soil coverage is a critical factor required to attain the above-mentioned agronomic advantages associated to mulching. PE mulches are little affected by the environment and ensure soil coverage. For biodegradable mulches, environmental agents (UV radiation, temperature, rainfall and humidity, wind and animal or plant activity) alone or in combination, may trigger the above-soil mulch degradation, cause breakings and diminish the elasticity of the mulch. Consequently, the mulch soil coverage is reduced.

A range of methods to determine mulch degradation in the field, based on qualitative estimation of breakings, on the mulch resistance to stress and, most frequently, on qualitative estimation of the mulch soil coverage, have been used [41]. The review of these methodologies has led to proposing a unified mulch field degradation scale [66]. Environmental factors depend on location, year and season; thus mulch repercussions are to be determined under a variety of conditions. Final results can vary substantially when biodegradable mulches are used by combining with different production technologies that change the environment. Deterioration (breakings, decrease in soil coverage) of two MB films and a WeedGuardPlus paper in a fresh market tomato crop was higher in the open field than in a high tunnel environment [59, 67]. Location also had an effect: mulch deterioration was ca. 60% greater in the open field than in high tunnel in one location, but ca. 150% greater in another open field with high winds, greater solar radiation and rainfall [59]. A very thick PLA nonwoven and PE mulches did not deteriorate under any of the conditions tested. Wortman et al. [49] also obtained higher soil coverage and less deterioration with nonwoven mulch fabrics than with biodegradable films.

An extensive study of mulch degradation in a processing tomato field with three biodegradable plastics (MB, BF, BFx), two papers (MimGreen, Saikraft) and PE has been carried out in five locations and three crop seasons under Mediterranean conditions [66]. The main factor responsible for degradation was found to be the material: PE did not show relevant degradation while BF and MB degraded more and faster than paper mulches. Among biodegradable mulches, BF degraded more than MB, and MB degraded more than BFx. Environmental factors triggering degradation, basically sun radiation, rainfall, and wind, and crop coverage, were identified as the second responsible degradation agents.

Although at the end of the crop season, soil coverage is less with black biodegradable mulches than with PE [66], the performance of biodegradable mulches inside tunnels [57, 68], or in the open field [35, 41, 42, 51–53, 56] shows that their lifespan is adequate for fresh market and processing tomato. The main agronomic effects remain basically unchanged when PE is substituted by most biodegradable plastic mulches. It has also to be noticed that the new generation of biodegradable films is more resistant to environmental deterioration than the initially available materials.

Not only the environment has an effect on mulch duration and on soil coverage, specific agronomic practices may also have an impact; among others, the time span between mulching and transplanting and the wetting of the mulch through irrigation. The age of the mulch material is also to be considered. Furthermore, mulch degradation is associated to the crop itself [66]. The readiness of the crop to cover the mulch, together with the physical contact between crop and mulch relates to the level of interaction between mulch and environment. Thus, it is not surprising that in the same location and season, soil coverage may be different for fresh market tomato and processing tomato crops; their crop production technologies are quite different.

Mulch and Microenvironment. The basis for the agronomic effects of mulches is that they modify plant and soil temperature and humidity mainly at early crop developmental stages. Once the crop has covered the mulch, microclimatic effects are not substantial. Most studies show that biodegradable materials increase soil temperature as much as PE films or slightly less. Candido et al. [57, 58] reported that black MB (12 and 15 µm) mulches increased maximum and minimum soil (sandy soil) temperature at 10 cm depth 1 °C less than black PE (50 µm).

A similar trend has been reported in other studies comparing black biodegradable mulches (MB, BF) with PE (15 µm; 25 µm) at 5 and 10 cm depth [52, 53, 59, 69]; soil temperature with paper mulches was intermediate between that with biodegradable mulches and bare soil. At 20 cm depth the effect of mulches on soil temperature was lower and at 30 cm they had little effect [58]. However, the use of a clear diffuse MB (40 µm) mulch resulted in 1 °C higher soil temperature than with black PE films at 20 cm depth [58]. Equally, Ngouajio et al. [35] registered higher soil temperature at 1 cm depth for Ecoflex (25–35 µm) than for PE (25 µm) films at the initials of the crop cycle. However, later on temperatures with biodegradable films dropped below those with PE films. Overall, black Ecoflex and PE mulches equivalently increased soil temperature. Also, although soil thermal amplitude (difference between daily maximum and minimum temperatures) under Mediterranean continental climate was not affected by mulching at 10 cm depth [53], it significantly increased at 5 cm depth [69]. Nonwoven mulches, SB-PLA [49, 59, 60], had limited effect on soil temperature; they tended to increase it less than PE or MB and Ecofilm mulches. Nonwoven mulch fabrics would be better adapted for crops growing in warm environments.

In hot climates, mainly at the early developmental stages when too warm temperatures would stress the plantlets, increase their sensibility to root diseases and ultimately severely compromise plantlet survival, the lower temperature increases caused by biodegradable as compared to PE films may be an advantage. On the other hand, the temperature of the black PE mulch surface may increase more than that in biodegradable mulches (MB), reaching thermal stress for young plants, over 60 °C. Under these conditions, which are frequent when transplanting is performed in late spring or early summer, the cost of repositioning transplant failures in PE mulched fields may economically justify the use of biodegradable mulches. On the contrary, in colder climates higher temperatures associated to PE films are an advantage.

Differences in soil humidity for biodegradable and conventional mulched crops have been hardly addressed. Martin-Closas et al. [50] found that PE and MB (18 μm) equally increased soil water potential during the cultivation cycle. However, water vapour transmission rate is considerably lower for PE than for biodegradable films: around 12 times for MB and BFx and 26 times for BF [42]. Although these differences in water transmission rate measured under laboratory conditions are relevant and imply higher water saving capacity for PE than for biodegradable films, this is not frequently apparent when soil humidity is measured in the field, nor are they found associated to higher yields. In the field, differences in soil humidity with different mulches have not been reported. The air relative humidity below the mulch offers an insight into the water exchange dynamics between soil and air for the different mulches. Usually, water saturation and condensation occurs earlier during the day for PE than for biodegradable mulches; the mean relative humidity below the PE mulch is also higher.

Regarding biodegradable nonwovens, Wortman et al. [49] found that a nonwoven (PLA) increased soil moisture in a tomato crop as much as plastic mulches (MB and Ecofilm). Overall, mulches increase and stabilize soil humidity in the field. According to water transmission properties, PE is expected to maintain higher soil water content than biodegradable mulches. In spite of that, indications of these differences between conventional and biodegradable films have not been recorded in the field.

4.2.2.2 Effects on Pepper and Aubergine

Pepper and aubergine are more demanding crops for temperature than tomato; thus, they are usually transplanted later in the season, from late spring to early summer. Mulching is frequently used to optimize the thermic environment of the crop and to advance transplanting; besides, among other advantages, it protects the crop from weeds. One of the main differences of pepper and aubergine crops from tomato is their erect grow pattern. Pepper develops smaller plants with narrower leaves than aubergine, which develops bigger plants with broader leaves. As a consequence, pepper is cultivated at higher plant densities than aubergine, which usually reaches maximum ground cover earlier than pepper.

Pepper was among the first crops where biodegradable mulching was evaluated. In Australia, the yield was found to be equivalent with recycled brown paper (100 g m^{-2}), with MB (black; 30 μm), and with a PE (white on black; 25 μm) mulch, and higher than in the unmulched weeded control [70]. The performance of the mulches was likely achieved by keeping the soil longer at a convenient temperature for root growth. At harvest, the buried area of the biodegradable mulch was significantly degraded, while the exposed area was basically intact. However, much thinner MB films had higher breakdown in a pepper crop than in a processing tomato crop [51]. The growing habit of pepper plants, erect and with thin vegetative growth, likely allowed higher interaction of the mulch with environmental factors than in tomato crops, thus facilitating the degradation of the mulch.

Equivalent pepper yields were obtained also in Australia by using MB grades NF803/P (12 and 15 μm), a new MB grade CF04/P (15 μm) and a PE mulch (25 μm) [46]. When the biodegradable mulches entered in contact with the fruits, they stuck to their surfaces. However, since fruits mostly develop without contact with the mulch, the problem was confined to a very limited percentage of them. This stickiness may affect quality in other species, since these fruits may rot when in contact with soil. Plant cultivar and seasonal conditions are likely to influence whether fruit set occurs near the soil or higher in the plant, and thus the subsequent risk of this problem to arise.

In Canada, mulches (MB 15 μm and PE 28 μm; black, clear or infrared–IR-transmitting) had little impact on yield and fruit quality and maturity of pepper, and no differences were found between them [45]. After 4 weeks in the field, and before weeds were established, the MB clear plastic started breaking and was disintegrated 4 weeks later. In contrast, the black MB mulch remained intact until the end of the crop season. In aubergine, both biodegradable and standard mulches equally increased yield and crop development. For slow growing erect crops, like pepper and aubergine, supplemental manual weed control was more relevant than for other more vigorous and sprawling crops [45].

The effects of spunbond, nonwoven PLA biodegradable mulch fabrics and biodegradable films (MB, 15 μm; Ecofilm, 19 μm) on soil coverage and macro-climate of a pepper crop were found to be similar to those on tomato shown previously. Biofabrics and biodegradable films did not increase pepper yield relative to bare soil [49]. However, the combination of warmer soil temperatures and high precipitation induced greater *Phytium* spp. infection in plants cultivated with biodegradable films than with biofabrics.

4.2.2.3 Effects on Cucurbit Crops

Mulching is commonly used in cucurbitaceous crops, like melon, watermelon, zucchini, pumpkin or cucumber. All these crops are spring–summer crops that need higher temperatures than tomato, and are very often cultivated in mild winter areas. In such areas, the main objective is to get the harvest to the market as soon as possible, when prices are highest. Clear mulches are the most suitable for this purpose because they provoke higher soil warming effect than black mulch. Black mulches are used when earliness is not the main goal. The earliness character of a material greatly depends on the transmissivity to solar and long IR radiation from soil.

The transmissivity in the long IR electromagnetic spectrum of a 20 μm MB black film (11%) was reported to be half of that of a 40 μm PE film (22%), in spite of being half the thickness [71]. That is to say, the potential for black MB mulch to maintain soil heat and to induce earliness, provided that the mulch remains intact, is higher than for PE. When equivalent clear films were tested, solar and long IR transmissivity were once again higher for the thicker PE than for the thinner MB film: 92 versus 87%, respectively, for solar radiation transmissivity, and 75 versus 24%, respectively, for long IR radiation transmissivity. Moreover, and in contrast to PE, MB films transfer solar radiation mostly by diffusion, which diminishes the efficiency of transmission. The earliness effect of the mulch materials depends on the radiation balance of the film, on the film integrity along the crop, and on the structure of the crop during the early development. Cucurbit crops usually cover the mulch much faster than tomato and other crops. As compared with tomato, canopy development proceeds faster and bigger plants are produced; also, they show sprawling growth habits.

Melon is the most frequent cucurbit crop. Candido et al. [72], in South Italy, were the first to study biodegradable mulching in melon. In the greenhouse, they reported similar yield and fruit quality with a clear PE film (50 µm) and with a diffusive clear MB film (25 µm). The average harvest was only one day shorter with PE than with MB mulches. The MB film was more efficient in rising soil temperatures, although the breaking of the film by weeds cancelled this effect. MB started degradation 30–40 days after set up, when the mulch function was mostly accomplished, and it was totally degraded at harvest (90–100 days).

In an open air trial in South-East Spain, two melon cultivars mulched with clear MB (18 µm) and conventional PE (25 µm) films produced equivalent yield and fruit sugar content [73], while black films performed worse. Thinner (15 µm) clear MB mulches allowed similar or higher yields than clear PE [74]. The degradation rate of the biodegradable mulches was adequate during the crop cycle, but scarce rainfall and low soil humidity favoured a considerable amount of plastic remaining in the soil [73]. In very dry regions soil microbiological activity is limited and the in-soil biodegradation is likely to be very slow. However, the potential to biodegrade remains; in a following trial, the biodegradable mulch was not visible after 5–6 months [74].

In accordance, increased crop yield and earliness were reported in Sicily for a winter melon crop [75] with clear PE (50 µm) and MB (18 µm) films as compared with equivalent black films. Vetrano et al. [76] reported that clear PE (50 µm) and black MB (20 µm) were alike for early and final production. However, under dry farm conditions also in Sicily, black biodegradable MB (15 µm) and PE (15 µm) mulches yielded similar early and final production but less than clear PE (15 µm) mulch [77]. Clear PE film yielded more than the black biodegradable mulch of equal thickness. Higher soil temperatures were associated with clear films, probably because of the higher solar radiation transmission rate. The melon production with mulching was double than that obtained without mulch; likely due to higher crop water availability. Interactions between mulch type and melon cultivar tested have also been described [76]: some cultivars but not others perform better with black MB than with clear PE.

When clear mulch films are used, pre-emergence herbicides or other techniques have to be implemented to avoid weeds emerging and breaking the mulch; which would buffer the mulch thermal effect on soil and reduce the effects on crop earliness. Green and brown PE mulches have provided the best combination of soil warming and weed control in strawberry [78]. In Spain and Italy, green MB mulches have proved to increase the mean and maximum soil temperature over black MB films (grade NF803, 18 and 15 μm respectively), and to a lesser extent over black PE films (25 and 50 μm respectively) [79, 80]. Under Mediterranean continental conditions in Spain, this led to 6 days' advancement of the first harvest, as compared with black MB film [79], but early production at 50% of the harvesting period was not affected. Final yield and quality were similar for the green and black MB and PE mulches, as was also reported in Portugal [81]. In the Spanish trial, a new grade of black MB mulch (CF04/P) was equivalent for earliness, total production and quality than the green MB film and black PE previously mentioned. Soil coverage along the crop cycle was higher with the new grade mulch than with the previous one (NF803/P) and the in-soil degradation was somewhat slower. In coastal Mediterranean climate, yield and quality increased with the green biodegradable films [80]. In Spain and Italy, degradation in the field was faster for the green than for the black film, but both of them maintained good soil coverage and weed control. After harvest the films disappeared with time when buried in the soil, but the process took longer for the green than for the black mulch. A preliminary respirometric test showed low incipient biodegradation rate for the green biodegradable film [81]; but the test was conducted with a soil low in organic matter content. The biodegradation rate of coloured biodegradable mulches with the corresponding functional additives may vary from that of the original materials; in the case of MB green films biodegradation slowed down.

Melon fruits may directly develop in contact with humid MB mulch that has started degrading; the film can stick to the fruit cover and decrease the commercial value. This was also evaluated by the two above-mentioned studies. In the Spanish dry continental climate only the green film adhered to the fruits, but the rate of fruits affected was not significant [79]. In the more humid coastal climate in Italy, the stickiness was higher, 6.8% for the black and 12% for the green mulch [80]. In Canada [45] equivalent marketable yield was obtained in a cantaloupe melon crop with biodegradable MB (15 μm) and with standard PE (28 μm) mulches. Clear films yielded higher than black ones.

Biodegradable mulches can be used in combination with other protected cultivation techniques. Under a nonwoven low tunnel (17 g m^{-2}; 25% shade) maintained the first month of the crop cycle, no differential effects on harvest date and total yield were found for PE (black, grey and brown; 50 μm) and biodegradable (grey; 12 μm) mulches, and weed control was equally successful with all of them [82].

Biodegradable mulches have been more limited tested with other cucurbit crops, but results are similar. In USA, different MB (grades NF0U/P and NF803/P; 12 and 15 μm) and PE films had no differential effect on watermelon yield and earliness [83]. In zucchini, Minuto et al. [68] found similar yields with MB (NF803/P; 12 and 15 μm) and with PE (50 μm) films, and Waterer [45] reports no consistent differences either among biodegradable (MB; 15 μm) and standard (PE; 28 μm) clear, black and wavelength selective mulches. In a cucumber crop in Poland, PLA and conventional PP nonwovens (50 g m^2 each) used as mulching equally increased yield as compared with bare soil [29]. The higher yields were not due to differences in soil temperature, but to the maintenance of soil moist. Also in cucumber in the USA, Wortman et al. [84] conducted field and high tunnel trials on performance and after use decomposition of two bioplastic films (MB, 15 μm; Ecofilm, 19 μm) and four experimental spunbond PLA nonwoven biofabrics of varying thickness. Effects on soil temperature, soil humidity, weed control, and mulch duration were parallel to those previously found for tomato. However, mulches had no effect on yield in cucumber. Since PLA is usually recalcitrant to degradation in the agricultural soil conditions, it was also unexpected that the relative rate of decay (mass of material recovered per soil surface) in the soil was equivalent for any of the biomulches. Also with cucumber, several mulch papers (brown, black, coated with black biodegradable film in one or two sides, wax coated, creped, etc.), biodegradable plastic mulches (MB, 18 and 30 μm), and especially mulches with a dark upper surface, increased yield in Finland. Dark-coloured mulches had the greatest soil warming effect and controlled weeds [85]. Black-printed paper mulch was suggested as an alternative to substitute for biodegradable film in cucumber production.

4.2.2.4 Effects on Strawberry, Lettuce and Other Crops

Strawberry annual production is usually in the open field or inside tunnels, for 8–12 months in rows with plastic mulch. It is considered a long-lasting crop. For this reason, thick films are used. The cultivation on elevated ridges covered with plastic avoids fruits contacting the wet soil, thus preventing the fruits to rot and optimizing quality and production. In a strawberry crop in Italy, yield was higher and earlier with MB (25–45 μm) than with PE (50 μm) mulch, and the duration of the MB film was adequate for the crop. Only 4% of the initial MB mulch weight remained into the soil one year after the mulch tillage [40]. In contrast, in Portugal, yield was greater with PE (40 μm) than with MB (20–30 μm) mulch and earliness was not affected by the type of mulch [86]. With the PE mulch, lower soil temperatures under the mulch initially in the crop cycle favoured the strawberry plant development, and more fruits per plant were developed. In open field and greenhouse trials in the same region of Portugal, Costa et al. [87] did not find differences in yield and soluble solids among MB (NF803/P, 18 μm and CF04/P; 18–20 μm) and PE (35 μm) mulch films.

Mulching is usual in lettuce, a short lasting crop, mainly for weed control and for improving quality and earliness. In the open field, equivalent yield and quality of a lettuce summer crop was found with black PE (25–50 μm) and MB (NF803/P; 12 and 15 μm) film mulches [68, 83]. In another greenhouse summer trial in South-East Spain, black and clear PE yielded more than biodegradable (15 μm) mulches [88]. Leaf development was bigger for PE cultivated plants. However, with MB mulch soil temperatures were higher. In summer, these temperatures may be too high and exceed the optimal thermal threshold for lettuce, thus limiting productivity.

Biodegradable mulching has also been evaluated in some crops which are not frequently mulched. In broccoli, cultivated as late summer–autumn or autumn–winter crop, both conventional and biodegradable mulches equally increased yield [83, 89, 90]. On sweet potato, a tropical crop that can be also grown in northern regions with 4–6 month frost-free season, mulches may be used to warm the soil and for weed control. In Korea, mulching sweet potato with PE and two biodegradable films (PBSA and PCL/Starch; 15–25 μm) resulted in higher or equivalent yields for the later [91]. In North-East USA, compared to bare soil, MB (15 μm) film increased total yield and reduced labour for weed control [92].

Clear mulch may be used to warm the soil and advance sweet corn maturity when planted in early spring. It is expected to improve seed germination and increase plant stand under cool soil conditions. However, emergence and stand count were not changed by using clear PE and MB films [45]. Both films advanced sweet corn maturity, but the mulches had no effect on yield. Finally, under very hot and dry summer conditions in Iran, black PE mulch yielded higher and controlled weeds better than a biodegradable mulch [93].

4.2.3   Effects on Fruit Crops and Vineyards

4.2.3.1   Specific Features of Fruit Crops and Vineyards

The characteristics of fruit crops have limited the use of plastic films. Fruit crops are mostly perennial: once planted, the orchard will stay for over 10 years, and soil labours after planting will be mostly restricted to row alleys. The main requirements for these orchard crops is long lasting films and lack of soil labours during the life of the orchard; thus, most plastic films can only be applied for special cases and in a few crops.

Also, the fruit plants go through a series of developmental stages (orchard establishment, juvenility, crop initiation, full crop production...) with different needs. Even more, planting is currently done using one or two years old 0.5–1.5 m high plants with extended root crowns, which make it difficult to use plastic films for mulching. In some cases smaller plants are planted (e.g. olive trees for super-intensive orchards, grapevines), but still the implementation of plastic films for mulching is difficult and the cost is high. Some species are planted with bare roots (apple, pear, peach, almond ...), some others with soil (olive...). To plant the trees, a hole in the soil has to be opened (ca. 0.5 m in diameter and 0.5–1 m deep) using an auger or a shovel/backhoe for low density orchards, or for higher density mechanized orchards deep furrows have to be opened.

By whatever means, special protection is needed for young trees, which are mostly grafted, during the first 2–3 years: they are highly sensitive to herbicides, are easily damaged by machinery used for controlling weeds, and require animal protection. In most cases, a trunk protection plastic is placed at planting; colour varying from black to white.

Biodegradable plastic films are currently shortly used in fruit production. Correspondingly, recent reviews of plastic and paper mulches [19, 24] have barely included fruit crops. The main focus to be presented is on the different uses of plastic films amenable to be replaced by biodegradable films, and on the main characteristics these materials have to fulfil. Available data on the use of biodegradable films are also exposed.

4.2.3.2   Mulching in Fruit Crops and Vineyards

In fruit orchards, control of weeds in the tree row is routinely accomplished by the use of herbicides or by cultivation, while in the alleyways, mowing complemented with some herbicide spraying is widely the practice. On rainfed conditions, the alleyways are also mechanically cultivated. In olive groves on rainfed conditions, plough or harrow parallel and perpendicular to tree rows is the common practice. A mulch-strip system that involves applications of fabrics, films or biomass in the tree row is only used in special occasions. Some advantages of the system are better water use efficiency and buffering soil temperatures. The drawbacks include increase of mice, voles and other rodents, higher incidence of some diseases (due to increased humidity on irrigated plots) and the barrier it builds that keeps out of rainwater (mostly required on rainfed conditions). Finally, the cost is higher and difficulties for establishment and maintenance may be encountered [94].

In organic fruit production no herbicides can be applied and mulching is used in the tree rows. Most of these mulches are organic: straw, bark, cover crops, almond husk or compost. They contribute to fertilization and can be supplied easily after plantation; shredded waste paper may be also used [95–102]. Film mulches are usually not implemented because installation and maintenance is difficult and costly.

The installation of the mulching film at planting is only possible for short young plants without leaves or branches (e.g. vineyards, strawberries); nonetheless, it is more complicated than for vegetable crops. For higher trees or for adult orchards, two pieces of the plastic film may be placed overlapping at both sides of the trees in some cases, hold in place by metal staples in the middle and with their outer edges buried in the soil [103]; alternatively, only one piece of plastic film may be used if transversally cut for each tree and covered with soil at the borders [104].

Neither biodegradable mulches, nor PE are used for orchards because of their short life. Mulching films should be installed at the orchard establishment and last at least for 3 years in good condition, or be replaced every year. However, for a multi-annual strawberry crop biodegradable mulches are mostly relevant for the first year. Thus, although the biodegradable plastic degrades during the first growing season, it may perform as well as low-density PE film on weed control and yield, and better than straw or paper wool [105].

Hostelter et al. [106, 107], using black and white geotextile, increased vine size and yield value, but not enough to compensate for the geotextile cost. Touchaleaume et al. [108] got better results in a vineyard; four black biodegradable films 40 μm-thick (Starch and PBAT: MB, new or with 10% recycled material, polypropylene carbonate (PPC) and PBAT and a blend of PLA and PBAT:Bioflex), improved the growth and the fruiting yield in a similar way to PE with respect to bare soil. Despite their short life span (5 months) the benefits of the biodegradable plastic films lasted for 2 years.

Alins et al. [104] tested two different systems of planting on an organic apple field with three black mulching films: 15 μm-thick low-density PE film, a 15 μm-thick biodegradable film made from starch and PBAT (MB), and 85 g/m^2 density paper film (MimGreen®). They cut back the young tree to 25 cm after planting before installing the mulch (Fig. 4.1) or made a perpendicular cut halfway to the plastic to circumvent the plant and then cover the cut area and the borders of the film with soil (Fig. 4.2). The first system strongly delayed plant growth (personal communication), and the second system did not provide satisfactory results for any of the films. All of them lost integrity after three months, presented cracks, and weeds emerged (Figs. 4.3 and 4.4). The paper film deteriorated faster: rain water accumulated over it softened the material, and winds cracked and removed the paper. Films used were thinner than those of Touchaleaume et al. [108], so they deteriorated faster.

Fig. 4.1 Mulching plastic film Mater-Bi® on an apple orchard where the trees were cut back to 25 cm after planting

Fig. 4.2 Mulching paper MimGreen® on an apple orchard where the paper was cut in order to circumvent the trees. Courtesy of Alins et al. [104]

Fig. 4.3 Biodegradable Mater-Bi mulching film 3 months after installation

Fig. 4.4 MimGreen paper mulching film of 3 months after installation

In the framework of the EU project LIFE ENV/IT/463 (BIOMASS), several experimental and demonstrative trials were carried out to stimulate the adoption of biodegradable plastics [68]. Crops included hazelnut trees, grapevines, Christmas trees (*Abies alba*), and perennial cultivation of strawberries. Trials consisted of fields mulched with long lasting 40 and 70 μm thick MB films and 50 μm thick black PE. Mater-Bi films provided good soil coverage for up to 6 months. After 14 months in the Christmas tree orchard, 70 μm MB produced better results than 40 μm MB, but lower than PE. After two years, the percentage of mulched soil was lower for both MB films than for the PE, and weeds were more abundant. With perennial cultivation of strawberries, degradation of the 40 μm MB started earlier, thus protection against weeds was shorter; while PE provided good protection during the 18 months trial. It is relevant to note that the Christmas trees trial was not irrigated, while the strawberry trial was drip irrigated.

After weed control, mulches are used for better water use. Water vapour permeability of the mulch needs to be low enough to reduce the moisture losses by evaporation, mainly in non-irrigated and dry areas [109]. On new planted non-irrigated raspberry and highbush blueberry multi-annual crops orchard, MB, 30 and 40 μm, has been found to improve water content, and increase temperature and vegetative growth, respect to bare soil [110].

For citrus plants in some wet areas, films impermeable to rain water are needed, so to allow less water go to the soil for higher quality fruits. The soil has to be fully covered by the film. In some occasions, the interest is for slow biodegradability of the film, which can help to control the moisture of the soil [111]. In regions with too much rain, for higher mandarin quality, impermeable films are needed to keep the soil dry. However, the soil should not be kept very dry for the entire period, and a biodegradable film that increases permeability with time as it degrades is a good alternative to irrigation under the film. Also, the degradation of the film helps to keep soil temperature lower. With less degraded film, soil temperature is kept too high. A combination of 30% PLA and 70% Ecoflex film gave better results in citrus fruits than 30% PLA with 60% Ecoflex and 10% modified starch [111].

A future mulching option to explore is the application of novel biodegradable polymeric mulching spray coatings. Biodegradable mulches can produce the coating directly in the field by spraying water solutions, thus covering the cultivated soil with the protective thin geomembrane produced. They are easy and cheap to implement in the field. Spray coatings are based on natural polysaccharides or on proteins [112–116].

4.3   Other Uses of Films on Fruit Crops and Vineyards

4.3.1   Trunk/Bark Protection

A common practice after planting young fruit trees is to protect them with a plastic cylinder around the bark, which is to enter into contact with the ground or, better, which is partially buried into the ground. The main objective of using this protection for the 2–3 first years after planting these trees, when the tree trunk is still sensitive to herbicides, is to allow the application of these chemicals to the tree row on hedgerow systems, or around the tree on low-density cropping systems. Also, the cylinders provide protection against rodents (e.g., rabbits, voles) and other mammals (e.g., deers, boars). Moreover, they reduce the damage to the trunk of the machinery used for weed control and of the stones they can throw. Some protection against sun burn, sun scald and winter frost is also provided. Some materials can also prevent lateral branches from growing. Trunk/bark protections introduce some interesting side effects. The temperature and the humidity inside the plastic area increase; higher temperatures hasten vegetative growth, but increased humidity can lead to higher disease occurrence.

Nowadays, a great variety of rigid (metal or plastic) or soft materials (plastic or geotextile) are used, and they can be reused several times. Since the guard tubes have to last at least 2–3 years in good conditions, no biodegradable film has been so far used. However, in small orchards, commercial paper wraps have been proposed as tree guards. They protect trees from weather damage, but not from animals or equipment; only when impregnated with repellent substances, the paper may also provide some protection against animals.

4.3.2   Containers for Planting Trees

Most perennial plants are planted with bare roots, at the dormant stage. Transport and management are easy and cheap, but some of them (olive and other evergreen tree or shrub species) are planted with soil and a container is needed. Nurseries have traditionally used rigid containers, but now they are changing to low-density PE film tubes that are cheaper and occupy less space. These pot-grown plants are grown at the nursery within the container, so that moving the trees is easier and they need less space than in the field. However, for transportation, they require more space than bare soil plants and they are heavier. At planting, the plant has to be separated from the container, so the plastic residue generated has to be collected and removed from the field. This presents an opportunity for biodegradable films: since the container with the tree could be planted together, reducing the management and cost.

Bilck et al. [117] showed that biodegradable films (poly-butylene adipate-co-terephthalate, PBAT) are a convenient alternative to low-density PE seedling bags for producing plant seedlings at nursery. The biodegradable bags remained intact for 60 days and, after being transplanted, they were completely biodegraded in 240 days. No differences were found between these plants and the plants transplanted without the bags. Also, pots made of MB that contained ornamental plants had no effects on plant quality and maintained original mechanical characteristics and physical properties—comparable to the ones of PP—during the whole cropping period, and during the marketing [118]. Tests on resistance to deformation and on resistance to crushing proved the quality of this product and the possibility for being used in highly mechanized cropping systems.

4.3.3   Fruit Protection Bags

Fruit protection bags are used to improve the microclimate around the fruit, for plant protection and for the combination of both. The bag is mainly waxed paper in a variety of colours depending on the main objective, but PE and PP can also be used [119–124]. Examples are:

- For late season peach, Paraffin paper bags are mainly used against flies (*Ceratitis capitata*), but also to avoid plant protection products to reach the fruit, to improve colour uniformity and size, to control diseases, and to change the microclimate around the fruit [125–127].
- For guava. A biodegradable plastic bag: a mixture of cassava starch and poly-butylene adipate-co-terephthalate) against insects and diseases [128].
- To improve the colour of red apples, such as 'Fuji' [129, 130]. Several coloured paper bags.

- For palm fruit. Craft paper bags increase temperature and humidity, thus promote early fruit ripening and quality [131].
- For kiwi fruit. Plastic and paper bags to alter temperature and humidity regimes cause different effects on fruit growth [132]. Similar results have been found for mango, red Chinese sand pears, litchi, Japanese persimmon, bananas and logan [119, 133–140].

The performance of the only reported case of biodegradable films, in guava, was similar to that of polypropylene bags, and the biodegradable bags presented the advantage that they could be left in the field after harvest [128].

4.3.4 Grafting Strips

Most of fruit trees are the result of joining together two plant materials: rootstock and scion, which is facilitated by grafting strips. Some different materials for the strips have been tested in order to protect the graft from dehydration and to hold it in place until the union is healed. Transparent PE tape is usually employed. Degradable tapes are another possibility, mixtures of a base of alkenes (polyolefins) with wax and rubber that degrades with the sunlight, like the BUDDY TAPE® (Kenogard). This is an elastic, self-adhesive, moisture proof and transparent tape that does not need to be cut and removed from the plant, saving time. Also, its elasticity keeps the union without air, preventing desiccation, which is highly important for tropical and some subtropical fruit trees. Thus, in rambutan trees [141], some citrus species [142], custard apple trees [143], and sapota trees [144], it has proved to be as efficient as the traditional PE tape; moreover, it has demonstrated to improve the length of the scion when under dry conditions. There are some patents of different types of graft strips.

One of the limitations for the use of this photodegradable grafting strip is lack of control over degradation, and so on holding. The time the graft strip is kept in place depends on lighting conditions. If it degrades too soon the graft will fail, and if it degrades too late the scion will not be able to sprout. Another limitation occurs if it does not fully biodegrade; a residue would be left on the field. Biodegradable grafting strips would start degradation in the air, mainly driven by light, and they would continue biodegrading when falling to the soil. An alternative for grafting some crops (vines) is the use of resin/wax [145, 146].

4.3.5 Reflecting Materials

In some orchards (apple, peach, plum, vineyard…) where light distribution to the lower part of the tree is important, reflecting films or geotextiles may be laid on the soil [103, 147–149]; they also reduce soil moisture [150]. They have to be strong to resist the passage of machinery for the last one–two months previous to the harvest. They have to be installed and removed every year, but they can be reused several years. However, the benefits may not pay for the cost. The advantage of biodegradable films is that there is no need to eliminate them from the ground after use, but on the other hand that increases the cost since they will not last for more than one year, so that new films will have to be laid every year.

An alternative to consider is the use of paintings (liquids films) that could be cheaper to apply to the soil. Blanke [147] used a biodegradable white paint that presented an initially large light reflexion, but which was washed off by the autumn rainfall, so that the quality of the fruits was poor. White paintings may offer better opportunities under drier conditions.

4.3.6   Holding Crowns and Hail Nets

The introduction of biodegradable films for holding crowns (peach trees) and hail nets is difficult since they are expected to last for the whole life of the orchard (10–15 years).

4.4   Soil Solarisation with Biodegradable Films

Soil solarisation is a well-known technique that has been applied for over 30 years. The aim of soil solarisation is to harness solar energy to raise the temperature of moistened soil. Covering the soil with clear PE or with another film is the most common mean of achieving solarisation [151]. The rise of soil temperature results in the elimination of most pathogenic fungi, bacteria and nematodes, and controls many seeds and seedlings from weeds; fortunately, beneficial soil organisms are able to either survive or to very efficiently recolonize the soil afterwards [152]. The potential of a film for solarisation is provided by high transparency to solar radiation (UV, PAR and NIR) and high impermeability to medium and far IR radiation [153]. The thinner the film is, the greater the soil heating is. Soil temperatures above 40 °C are usually considered as detrimental for soil diseases and for weeds.

Solarisation has been mainly used for intensive field and greenhouse vegetable production of strawberries, solanaceous, cucurbits and lettuce crops, mostly in locations with high summer radiation. The beneficial effects of soil solarisation on crop development and yield are not only due to better controlling crop predators and competitors, but also to provoking soil chemical and physical changes that end up in higher plant nutrient availability. This later effect is more important in organic cropping systems, where the nutrients are supplied only through the remains of crops, compost or manure. Higher soil temperatures increase the mineralization of organic matter, therefore accelerating the release of nutrients. As mentioned above, one of the main limitations of the techniques based on plastic films is their disposal after use; thus biodegradable plastic sheets are to be a good alternative to conventional films.

In the late 90s, based on spectra–radiometric characteristics, Manera et al. [9] suggested the suitability of a biodegradable film (MB/Z) for soil solarisation. They firstly [153] evaluated a clear biodegradable film, made of polyester amide (probably BAK 1095), for solarisation of a greenhouse soil (sandy clay), as compared to an EVA and to a coextruded (PE-EVA-PE) film. Despite the initial convenient performance of the biodegradable film on temperature, it had a short lifetime and started to degrade too early, limiting the in-soil thermal effect in time. Nevertheless, the control on the nematode *Meloidogyne incognita* was as effective as that of other two more persistent tested films. The low initial concentration of nematodes and the early soil warming effect of the biodegradable film, accounted for satisfactory control.

More recently, in a soil (sandy loam) with a higher degree of nematode (*Meloidogyne javanica*) infestation, Scopa et al. [154] reported lower nematicide effectiveness for biodegradable than for EVA and coextruded films, which was associated to the lower soil heating capability of the former ones. When biodegradable films were used, the yield of succeeding tomato and melon crops was over that of the corresponding non-solarised crops, but significantly lower than when solarizing with the other films. Main agronomic and environmental differences among polyester amide biodegradable film and conventional films were associated to the former earlier degradation, which started from day 15 after laying on.

Castronuovo et al. [155] compared the soil solarising effects of a starch based translucent biodegradable film (MB) to clear PE and EVA films, both in the greenhouse and in the open field. The biodegradable film transmittance to the UV-PAR-NIR radiation was lower, but it exhibited the best optical performance in the IR region. The solarising effect of all tested materials allowed an equivalent control of the nematode *M. javanica* and weeds (except *C. rotundus*); yield and fruit quality of the subsequent melon crop were also equivalent both in the greenhouse and in the open field. Similar results were found by Candido et al. [156] when comparing a MB to an EVA film in a sandy greenhouse soil. The soil thermal effect during solarisation was equal for both materials, and the level of control of *M. javanica* and the yields achieved in a subsequent eggplant crop were also equivalent. Further studies with MB, LDPE and EVA films for solarising greenhouse soil found equivalent and positive agronomic effects (production, quality and soil nematode control) of the films on melon and lettuce crops [58].

In contrast, in open field experiments [157], an EVA and a photoselective (Polydak) film, but not a biodegradable film (MB), produced soil sterilization. Transmittance in the medium IR was lower for the MB film, but the severity of fungus diseases was higher. Yields in cucumber, watermelon and marrow crops following solarisation were higher with the EVA and the photoselective film than with the biodegradable one; only in melon were equivalent. It can be speculated that the fast degradation and breakdown of the MB film in the open field prevents obtaining thermal effects for solarising equivalent to those of conventional films. For an effective use of MB in solarisation, the duration of the film and the mechanical properties should be improved.

Bonanomi et al. [158] investigated the impact of solarisation with MB and PE (polysolar) films in clay and sandy soils and on tomato and lettuce crops. They found similar soil warming effects with both materials, but the biodegradable film only allowed a short period of solarisation (less than 1 month), while two months are usually required for completing the treatment. Plant mortality was reduced in both types of soils in the lettuce crop, with no differences between materials; however, and only in the clay soil, weeds were more persistent with the biodegradable than with the PE film.

Candido et al. [159, 160] analysed solarisation on weed control and yield in two lettuce crops (fall–winter; spring) under field and greenhouse conditions. As reported previously, MB film was completely torn in the field and started degrading in the greenhouse 15 days after laying on. Soil temperatures over 40 °C lasted longer in the greenhouse than in the field, and under PE, EVA or coextruded PE-EVA films than under the biodegradable material. However, both in the greenhouse and in the field, weed density and biomass were strongly reduced by soil solarisation, without differences among the tested materials. Most annual weed species were completely controlled. Only *Amaranthus* spp. and *P. oleracea* were poorly controlled or even stimulated with the biodegradable film, very likely due to the limited thermal effect of the material [159]. Perennial weeds were not controlled in any case. Solarized soils produced higher yields, without differences among tested materials, but crop duration was shortened only in the greenhouse. Quality of field lettuce was analysed and found to be unaffected by solarisation [160].

Strawberry crop is very sensitive to soil borne diseases, *Rhizoctonia solani*, *Sclerotium rolfsii* and *Fusarium oxysporum* f. sp. *fragariae* being the most relevant ones. Raj [161] found that the efficiency in heating the soil and in decreasing the viability of the three pathogen propagules of MB for soil solarisation was slightly less than for PE films; however, soil solarisation was satisfactory with any of the two materials.

Solarisation with biodegradable films has been basically tested with MB, and to some extent also with polyester amine (BAK 1095), which is not anymore available in the market. Experimental results with MB have been mostly produced between 1999 and 2005. Only Raj [161] has published later experimental work from 2008 to 2009. Both films proved to have good solarisation potential, although future improvements in their duration in the open field and in increasing transmittance to solar radiation are expected to significantly increase effectiveness. Based on the above results, MB can be counselled for solarisation, preferably for greenhouse soils and for organic farming. Its effectiveness in the open field is to be dependent on the early degradation it may undergo, mainly in locations with high solar radiation. However, the present MB film, which has a different grade, and other biodegradable materials, remain to be tested for their solarisation potential.

4.5   Future Prospects and Perspectives

The main current application of biodegradable films in agriculture is mulching. After over 15 years of agronomic research, biodegradable mulching has demonstrated the potential to substitute PE in a wide diversity of crops, but its adoption is still limited. The main barrier remaining is the higher price of biodegradable materials, versus cheaper PE. The overall cost is lessened when considering the cost of removal and disposal to the waste manager. Still the cost is not balanced with PE, although this depends on the specific PE mulch used in every region. However, the weakness in the economic evaluation currently considered is the omission of including the cost of recycling PE mulches. Recycling is performed out of the agricultural sector and because of that the farmer takes no responsibility in the waste. Even when this is the case, the producer of the waste still remains responsible for the cost of the waste management. When the cost of recycling is considered, balanced costs for PE and biodegradable mulches are likely to be encountered. In the present situation, the cost is transferred into an environmental cost, which arises from a wrong management of the waste, and which is still more difficult to evaluate. At this level, governments are expected to decide whether take responsibility to maintain an input that produces a waste difficult to manage or to directly or indirectly facilitate the use of more friendly materials. At the same time, manufacturers are expected to optimize the materials together with the processes used to produce them. Products are to become more competitive and to overcome some constraints presently remaining, such as the low biobased content of commercial materials, the limited transparency and the weakness of clear films, while they have to improve permeability to gases focused on specific applications, enhance duration of the films for long lasting crops, and avoid competition with the food production sector.

A second barrier for the implementation of biodegradable materials in the field is the insufficient training provided to users. Training about laying on the films, information on breakdown of the materials, and in-soil biodegradation in the diversity of crops and conditions is required. Field demonstration and extension activities are to be developed together with basic and applied research. Finally, the availability of biodegradable materials within the market is the third barrier. On this aspect the situation has been consistently improved along the years, both in the diversity of materials and in their conversion and distribution.

Although vegetable mulching is the main present application of biodegradable films, in the near future they are likely to be implemented for other agricultural applications, mainly in fruit crop production and in extensive crops. Also, they may facilitate the agricultural exploitation of areas where production is limited by environmental constraints (water scarcity, soil erosion, etc.). Meanwhile, the use of biodegradable materials keeps growing, both in organic farming and in conventional fields. Users are increasingly aware of their responsibility in front of the environment and they show interest in further learning [162].

References

1. Martin-Closas L, Rojo F, Picuno P et al (2009) Potential of biodegradable materials for mulching as an alternative to polyethylene. Paper presented at the XVIII International Congress Plastics in Agriculture. Almeria, 23–25 Nov 2009
2. Levitan L, Barros A (2003) Recycling agricultural plastics in New York State. Environmental Risk Analysis Program, Cornell University, Ithaca. http://cwmi.css.cornell. edu/recyclingagplastics.pdf. Accessed 20 Jan 2016
3. Steinmetz Z, Wollmann C, Schaefer M et al (2016) Plastic mulching in agriculture. Trading short-term agronomic benefits for long-term soil degradation? Sci Total Environ 550: 690–705
4. Kolybaba M, Tabil LG, Panigrahi S, et al (2003) Biodegradable polymers: past, present, and future. Paper nr: RRV03-0007 presented at the CSAE/ASAE annual intersectional meeting, Fargo, 3–4 Oct, pp 1–15
5. Hussain I, Hamid H (2003) Plastics in agriculture. In: Andrady AL (ed) Plastics and environment. Wiley, Hoboken, pp 185–209
6. Martin-Closas L, Pelacho AM (2011) Agronomic potential of biopolymer films. In: Plackett D (ed) Biopolymers. New materials for sustainable films and coatings. Wiley, Chichester, pp 277–299
7. Weber C (1998) Mulching with biologically degradable plastic. Mulchen mit biologisch abbaubarer Folie. KTBL-Arbeitspapier 251:78–83
8. Nagata M, Tadeo BD, Hiyoshi K et al (1998) Biodegradable film mulching cultivation system for early season culture rice: thermal image analyses of mulch films using thermography (part I). Bull Fac Agric Miyazaki University 45:143–152
9. Manera C, Margiotta S, Picuno P (1999) Innovative/biodegradable plastic films for protected cultivation. Colture Protette 28:59–64
10. Groot L, Pruschke K, Schüsseler P et al (2000) Biologisch abbaubare Werkstoffe im Gartenbau. KTBL-Schrift 386, Darmstadt
11. Kyrikou I, Briassoulis D (2007) Biodegradation of agricultural plastic films: a critical review. J Polym Environ 15:125–150
12. Hayes DG, Daharmalingam S, Wadsworth LC et al (2012) Biodegradable agricultural mulches derived from biopolymers. In: Khemani K et al (eds) Degradable polymers and materials: principles and practices, 2nd edn. ACS symposium series, ACS, Washington, pp 201–223
13. JBPA (2016) Japan BioPlastics Association. GreenPla products. http://www.jbpaweb.net. Accessed 20 Jan 2016
14. Tachibana Y, Masuda T, Funabashi M et al (2010) Chemical synthesis of fully biomass-based poly(butylene succinate) from inedible-biomass-based furfural and evaluation of its biomass carbon ratio. Biomacromolec 11:2760–2765
15. Vinçotte (2016) http://www.okcompost.be. Accessed 20 Jan 2016
16. BPI (2016) Biodegradable Products Institute. http://www.bpiworld.org. Accessed 20 Jan 2016
17. Din Certco (2016) http://www.dincertco.de. Accessed 20 Jan 2016
18. Heller B, Starke V, Straeter C et al (2008) Biologisch abbaubare Mulchfolien aus nachwachsenden Rohstoffen. Informationen und Verwendungshinweise. FBAW. Forschungsanstalt für Gartenbau Weihenstephan, Hannover
19. Kasirajan S, Ngouajio M (2012) Polyethylene and biodegradable mulches for agricultural application: a review. Agron Sustain Dev 32:501–529
20. Feuilloley P, Cesar G, Benguigui L et al (2005) Degradation of polyethylene designed for agricultural purposes. J Polym Environ 13:349–355
21. Selke S, Auras R, Nguyen TA et al (2015) Evaluation of biodegradation-promoting additives for plastics. Environ Sci Technol 49:3769–3777

22. Briassoulis D, Babou E, Hiskakis et al (2015) Analysis of long-term degradation behaviour of polyethylene mulching films with pro-oxidants under real cultivation and soil burial conditions. Environ Sci Pollut Res 22:2584–2598
23. Briassoulis D, Babou E, Hiskakis M et al (2015) Degradation in soil behaviour of artificially aged polyethylene films with pro-oxidants. J Appl Polym Sci 132:42289
24. Haapala T, Palonen P, Korpela A et al (2014) Feasibility of paper mulches in crop production—a review. Agric Food Sci 23:60–79
25. Liakatas A, Clark JA, Montheith JL (1986) Measurements of heat-balance under plastic mulches. Part1. Radiation balance and soil heat-flux. Agric Forest Meteorol 36:227–239
26. Tarara JM (2000) Microclimate modification with plastic mulch. HortScience 35:169–180
27. Smith BR, Wadsworth LC, Kamath MG et al (2008) Development of next generation biodegradable mulch nonwovens to replace polyethylene plastic. Paper resented at International Conference on Sustainable Textiles (ICST 08), Wuxi, China, 21–24 Oct 2008
28. Siwek P, Libik A, Gryza I et al (2009) Physico-mechanical properties and utilities of melt-blown biodegradable nonwovens. Paper presented at the XVIII International Congress on Plastics in Agriculture. Almeria, 23–25 Nov 2009
29. Siwek P, Domagala-Swiatkiewicz I, Kalisci A (2015) The influence of degradable polymer mulches on soil properties and cucumber yield. Agrochimica 59:108–123
30. Otey FH, Westhoff RP, Russell CR (1975) Starch-based plastics and films. In: Proceedings of technical symposium: nonwoven product technology international nonwovens disposables association, Miami Beach, Florida, Mar 1975, pp 40–47
31. Swanson CL, Westhoff RP, Doane WM et al (1987) Starch-based blown films for agricultural mulch. Div Polym Chem 28:105–106 (ACS, Polymer Preprints)
32. Lichocik M, Krucińska I, Ciechańska D et al (2012) Investigations into the impact of the formation temperature on the properties of spun-bonded nonwovens manufactured from PBSA. Fibers Text East Eur 20:70–76
33. Sztanjnowski S, Krucińska I, Sulak K et al (2012) Effects of the artificial weathering of biodegradable spun-bonded PLA nonwovens in respect to their application in agriculture. Fibers Text East Eur 20:89–95
34. Zawiska I, Siwek P (2014) The effects of PLA biodegradable and polypropylene nonwoven crop mulches on selected components of tomato grown in the field. Folia Hort 26:163–167
35. Ngouajio1 M, Auras R, Fernández RT et al (2008) Field performance of aliphatic-aromatic copolyester biodegradable mulch films in a fresh market tomato production system. HortTechnology 18:605–610
36. Tadeo BD, Nagata M, Mitaria M et al (2000) Study on mechanization of mulching cultivation using biodegradable film for early season culture rice (Part 1). J JSAM 62: 144–153
37. Briassoulis D (2004) An overview on the mechanical behaviour of biodegradable agricultural films. J Polym Environ 12:65–81
38. Briassoulis D (2004) Mechanical design requirements for low tunnel biodegradable and conventional film. Biosyst Eng 87:209–223
39. Briassoulis D (2006) Mechanical behaviour of biodegradable agricultural films under real field conditions. Polym Degrad Stab 91:1256–1272
40. Scarascia-Mugnozza G, Schettini E, Vox G et al (2006) Mechanical properties decay and morphological behaviour of biodegradable films for agricultural mulching in real scale experiment. Polym Degrad Stab 91:2801–2808
41. Martin-Closas L, Pelacho AM (2004) Biodegradable mulches as alternative to paper and polyethylene mulches in an organic production system. Paper presented at the VI Congress SEAE. Almeria, 27 Sept–2 Oct 2004
42. Martin-Closas L, Picuno P, Pelacho AM et al (2008) Properties of new biodegradable plastics for mulching, and characterization of their degradation in the laboratory and in the field. Acta Hortic 801:275–282
43. Kijchavengkul T, Auras R, Rubino M et al (2008) Assessment of aliphatic–aromatic copolyester biodegradable mulch films. Part I: field study. Chemosphere 71:942–953

44. Martin-Closas L, Bach A, Pelacho AM et al (2008) Biodegradable mulching in an organic tomato production system. Acta Hortic 767:267–274
45. Waterer D (2010) Evaluation of biodegradable mulches for production of warm-season vegetable crops. Can J Plant Sci 90:737–743
46. Limpus S, Heisswolf S, Kreymborg D et al (2012) Comparison of biodegradable mulch products to polyethylene in irrigated vegetable, tomato and melon crops. Department of Agriculture, Fisheries and Forestry. Final report Project MT09068. Horticulture Australia Ltd., Sidney
47. Novamont (2016) Biodegradable and compostable mulch film. User manual. http://www.duboisag.com/en/biodegradable-compostable-black-mulch-film-bio360.html. Accessed 14 May 2016
48. Asobiocom (2015) Plástico biodegradable para acolchado. Manual de uso. http://www.asobiocom.es. Accessed 14 May 2016
49. Wortman SE, Kadoma I, Crandall MD (2015) Assessing the potential for spun-bond, nonwoven biodegradable fabric as mulches for tomato and bell pepper crops. Sci Hortic 193:209–217
50. Martin-Closas L, Soler J, Pelacho AM (2003) Effect of different biodegradable mulch materials on an organic tomato production system. In: Biodegradable materials and fiber composites in agriculture and horticulture. KTBL-Schrift 414, Darmstadt, pp 78–85
51. Armendariz R, Macua JI, Lahoz I et al (2006) The use of different plastic mulches on processing tomatoes. Acta Hortic 724:199–202
52. Moreno MM, Moreno A (2008) Effect of different biodegradable and polyethylene mulches on soil properties and production in a tomato crop. Sci Hortic 116:256–263
53. Moreno MM, Moreno A, Mancebo I (2009) Comparison of different mulch materials in a tomato (Solanum lycopersicum L.) crop. Span J Agric Res 7:454–464
54. Anzalone A, Cirujeda A, Aibar J et al (2010) Effect of biodegradable mulch materials on weed control in processing tomatoes. Weed Technol 24:369–377
55. Cirujeda A, Aibar J, Anzalone A et al (2012) Biodegradable mulch instead of polyethylene for weed control of processing tomato production. Agron Sustain Dev 32:889–897
56. Macua JI, Jiménez E, Suso ML et al (2013) The future of processing tomato crops in the Ebro valley lies with the use of biodegradable mulching. Acta Hortic 971:143–146
57. Candido V, Miccolis V, Castronuovo D et al (2006) Mulching studies in greenhouse by using eco-compatible plastic films on fresh tomato crop. Acta Hortic 710:415–420
58. Candido V, Miccolis V, Castronuovo D et al (2007) Eco-compatible plastic films for crop mulching and soil solarisation in greenhouse. Acta Hortic 761:513–519
59. Miles CA, Wallace R, Wszelaki A et al (2012) Deterioration of potentially biodegradable alternatives to black plastic mulch in three tomato production regions. HortScience 47:1270–1277
60. Cowan JS, Miles CA, Andrews PK et al (2014) Biodegradable mulch performed comparably to polyethylene in high tunnel tomato (Solanum lycopersicum L.) production. J Sci Food Agric 94:1854–1864
61. Cirujeda A, Anzalone A, Aibar J et al (2012) Purple nutsedge (Cyperus rotundus L.) control with paper mulch in processing tomato. Crop Prot 39:66–71
62. Moreno C, Mancebo I, Tarquis AM et al (2014) Univariate and multivariate analysis on processing tomato quality under different mulches. Sci Agric 71:114–119
63. Shogren RL, Hochmuth RC (2004) Field evaluation of watermelon grown on paper-polymerized vegetable oil mulches. HortScience 39:1588–1591
64. Cirujeda A, Aibar J, Moreno MM et al (2013) Effective mechanical weed control in processing tomato: seven years of results. Renew Agric Food Syst 30:223–232
65. Fontanelli L, Raffaelli M, Martelloni L et al (2013) The influence of non-living mulch, mechanical and thermal treatments on weed population and yield of rainfed fresh-market tomato (Solanum lycopersicum L.). Span J Agric Res 11:593–602
66. Martín-Closas L, Costa J, Cirujeda A et al (2016) Above-soil and in-soil degradation of oxo- and bio-degradable mulches: a qualitative approach. Soil Res 54:225–236

67. Cowan JS, Saxton AM, Liu H (2016) Visual assessments of biodegradable mulch deterioration are not indicative of changes in mechanical properties. HortScience 51:245–254

68. Minuto G, Pisi L, Tinivella F et al (2008) Weed control with biodegradable mulch in vegetable crops. Acta Hortic 801:291–297

69. Moreno MM, Cirujeda A, Aibar J et al (2016) Soil thermal and productive responses of biodegradable mulch materials in a processing tomato (*Lycopersicon esculentum* Mill.) crop. Soil Res 54:207–215

70. Olsen JK, Gounder RK (2001) Alternatives to polyethylene mulch film, a field assessment of transported materials in capsicum (*Capsicum annuum* L.). Aust J Exp Agric 41:93–103

71. Vox G, Schettini E, Scarascia-Mugnozza G (2005) Radiometric properties of biodegradable films for horticultural protected cultivation. Acta Hortic 691:575–582

72. Candido V, Miccolis V, Gatta G et al (2003) Innovative films for melon mulching in protected cultivation. Acta Hortic 614:379–386

73. Gonzalez A, Fernandez JA, Martin P et al (2003) Behaviour of biodegradable film for mulching in open-air melon cultivation in South-East Spain. Biodegradable materials and fiber composites in agriculture and horticulture. KTBL-Schrift, Darmstadt, pp 71–77

74. Lopez J, Gonzalez A, Fernandez JA et al (2007) Behaviour of biodegradable films used for mulching in melon cultivation. Acta Hortic 747:125–130

75. Incalcaterra G, Sciortino A, Vetrano F et al (2004) Agronomic response of winter melon (*Cucumis melo inodorus* Naud.) to biodegradable and polyethylene film mulches, and to different planting densities. Options Mediterraneennes 60:181–184

76. Vetrano F, Fascella S, Iapichino G et al (2009) Response of melon genotypes to polyethylene and biodegradable starch-based mulching films used for fruit production in the Western coast of Sicily. Acta Hortic 807:109–113

77. Iapichino G, Mustazza G, Sabatino L et al (2014) Polyethylene and biodegradable starch-based mulching films positively affect winter melon production in Sicily. Acta Hortic 1015:225–231

78. Johnson MS, Fennimore SA (2005) Weed and crop response to coloured plastic mulches in strawberry production. HortScience 40:1371–1375

79. Martin-Closas L, Tura J, Rojo F et al (2010) Agronomic evaluation of a new generation of biodegradable mulch films of Mater-Bi[®] in a melon crop. In: Science and horticulture for people. 28th international horticultural Congress. August Lisboa 2010. Abstracts, vol II, p 617

80. Filippi F, Magnani G, Guerrini S et al (2011) Agronomical evaluation of green biodegradable mulch on melon. Ital J Agron 6(e18):111–116

81. Saraiva A, Costa R, Carvalho L et al (2012) The use of biodegradable mulch films in muskmelon crop production. Basic Res J Agric Sci Rev 1(4):88–95

82. Benincasa P, Massoli A, Polegri L et al (2014) Optimising the use of plastic protective covers in field grown melon on a farm scale. Ital J Agron 9:8–14

83. Miles C, Klingler E, Nelson L et al (2007) Alternatives to plastic mulch in vegetable production systems. http://vegetables.wsu.edu/MulchReport07.pdf. Accessed 4 Sept 2016

84. Wortman SE, Kadoma I, Crandall MD (2016) Biodegradable plastic and fabric mulch performance in field and high tunnel. HortTechnology 26:148–155

85. Haapala T, Palonen P, Tamminen A et al (2015) Effects of different paper mulches on soil temperature and yield of cucumber (*Cucumis sativus* L.) in the temperate zone. Agric Food Sci 24:52–58

86. Andrade CS, Palha MG, Duartec E (2014) Biodegradable mulch films performance for autumn-winter strawberry production. J Berry Res 4:193–202

87. Costa R, Saraiva A, Carvalho L et al (2014) The use of biodegradable mulch films on strawberry crop in Portugal. Sci Hortic 173:65–70

88. Lopez-Marin J, Abrusci C, Gonzalez A (2012) Study of degradable materials for soil mulching in greenhouse-grown lettuce. Acta Hortic 952:393–398

89. Lopez-Marin J, Gonzalez A, Fernandez JA et al (2012) Biodegradable mulch film in a broccoli production system. Acta Hortic 933:439–444
90. Cowan JS (2013) The use of biodegradable mulch for tomato and broccoli production: crop yield and quality, mulch deterioration and growers perception. Ph.D. dissertation, Washington State University, Pulmann
91. Lee JS, Jeong KH, Kim HS et al (2009) Bio-degradable mulching for sweetpotato cultivation. Korean J Crop Sci 54:135–142
92. Sideman RG (2015) Performance of sweetpotato cultivars grown using biodegradable black plastic mulch in New Hampshire. HortTechnology 25:412–416
93. Rajablarijani HR, Mirshekari B (2014) Sweet corn weed control and yields in response to sowing date and cropping systems. HortScience 49:289–293
94. Zibri W (2013) Efectos del acolchado sobre distintos parámetros del suelo y de la nectarina en riego por goteo. Ph.D. dissertation, University of Lleida, Spain
95. Guerra B, Steenwerth K (2012) Influence of floor management technique on grapevine growth, disease pressure, and juice and wine composition: a review. Am J Enol Vitic 63:149–164
96. Hogue EJ, Cline JA, Neilsen G et al (2010) Growth and yield responses to mulches and cover crops under low potassium conditions in drip-irrigated apple orchards on coarse soils. HortScience 45:1866–1871
97. Jacometti M, Wratten S, Walter M (2007) Management of understorey to reduce the primary inoculum of *Botrytis cinerea*: enhancing ecosystem services in vineyards. Biol Control 40:57–64
98. Lordan J, Pascual M, Villar JM et al (2015) Use of organic mulch to enhance water-use efficiency and peach production under limiting soil conditions in a three-year-old orchard. Span J Agric Res 13:1–9
99. Merwin IA, Rosenberger DA, Engle CA et al (1995) Comparing mulches, herbicides, and cultivation as orchard groundcover management systems. HortTechnology 5:151–158
100. Neilsen G, Neilsen D, Hogue E et al (2013) Soil management in organic orchard production systems. Acta Hortic 1001:347–356
101. Niggli U, Weibel FP, Potter CA (1989) Unkrautbekämpfung mit organischen Bodenbedeckungen in Apfelanlagen: Auswirkungen auf Ertrag, Fruchtqualität und Dynamik des Stickstoffs in Bodenlösung. Gartenbauwissenschaft 54:224–232
102. Verdu A, Mas M (2007) Mulching as an alternative technique for weed management in mandarin orchard tree rows. Agron Sustain Dev 27:367–375
103. Sandler HA, Brock PE, Heuvel JEV (2009) Effects of three reflective mulches on yield and fruit composition of Coastal New England winegrapes. Am J Enol Vitic 60:332–338
104. Alins G, Alegre S, Martin-Closas L (2012) Avaluació d'encoixinats per a control de males herbes en fruiters. VI Jornada Fructicultura ecològica, IRTA-University of Lleida
105. Daugaard H (2008) The effect of mulching materials on yield and berry quality in organic strawberry production. Biol Agric Hortic 26:139–147
106. Hostetler GL, Merwin IA, Brown MG et al (2007) Influence of geotextile mulches on canopy microclimate, yield, and fruit composition of cabernet franc. Am J Enol Vitic 58:431–442
107. Hostetler GL, Merwin IA, Brown MG et al (2007) Influence of undervine floor management on weed competition, vine nutrition, and yields of pinot noir. Am J Enol Vitic 58:421–430
108. Touchaleaume F, Martin-Closas L, Angellier-Coussy H et al (2016) Performance and environmental impact of biodegradable polymers as agricultural mulching films. Chemosphere 144:433–439
109. Hegazi A (2000) Plastic mulching for weed control and water economy in vineyards. Acta Hortic 536:245–250
110. Girgenti V, Peano C, Giuggioli N et al (2012) First results of biodegradable mulching on small berry fruits. Acta Hortic 926:571–576

111. Tachibana Y, Maeda T, Ito O et al (2009) Utilization of a biodegradable mulch sheet produced from poly(lactic acid)/Ecoflex®/modified starch in mandarin orange groves. Int J Mol Sci 10:3599–3615

112. Immirzi B, Santagata G, Pace Ed et al (2008) Preparation and characterization of cellulosic fibers and chitosan composites for agricultural application. In: Proceedings of the polymer processing society 24th annual meeting, Salerno, Italy 15–19 June 2008, pp P-032

113. Mormile P, Petti L, Rippa M et al (2007) Monitoring of the degradation dynamics of agricultural films by IR thermography. Polym Degrad Stabil 92:777–784

114. Schettini E, Vox G, De Lucia B (2007) Effects of the radiometric properties of innovative biodegradable mulching materials on snapdragon cultivation. Sci Hortic 112:456–461

115. Schettini E, Sartore L, Barbaglio M et al (2012) Hydrolyzed protein based materials for biodegradable spray mulching coatings. Acta Hortic 952:359–366

116. Malinconico M, Immirzi B, Santagata G et al (2008) An overview on innovative biodegradable materials for agricultural applications. In: Moeller, HW (ed) Progress in polymer degradation and stability research, Nova Science Publishers, Inc., pp 69–114

117. Bilck AP, Olivato JB, Yamashita F et al (2014) Biodegradable bags for the production of plant seedlings. Polimeros-Ciencia e Tecnologia 24:547–553

118. Minuto G, Minuto A, Pisi L et al (2008) Use of compostable pots for potted ornamental plants production. Acta Hortic 801:367–372

119. Biasi LA, Peressuti RA, Telles CA et al (2007) Qualidade de frutos de caqui "iro"ensacados com diferentes embalagens. Semina-Ciencias Agrarias 28:213–217

120. Coelho LR, Leonel S, Crocomo WB et al (2008) Avaliaçao de diferentes materiais no ensacamento de pêssegos. Rev Bras Frutic 30:822–826

121. Lipp JP, Secchi VA (2002) Ensacamento de frutos: uma antiga prática ecológica para controle da mosca-das-frutas. Agroecologia e Desenvolvimento Rural Sustentável 3:53–58

122. Malgarim BM, Mendes CDP (2007) Ensacamento de goiabas visando o manejo ecologico de moscas-das-frutas. Rev Bras Agroecologia 2:706–707

123. Mazaro S, Gouvea A, Citadin I et al (2005) Ensacamento de figos cv. "oxo de Valinhos". Scientia Agraria 6:59–63

124. Santos J, Wamser A, Denardi F (2007) Qualidade de frutos ensacados em diferentes genotipos de macieira. Ciencia Rural 37:1614–1620

125. Jia HJ, Araki A, Okamoto G (2005) Influence of fruit bagging on aroma volatiles and skin coloration of 'Hakuho' peach (*Prunus persica* Batsch). Postharvest Biol Tech 35:61–68

126. Kim Y, Kim S, Park J et al (2003) Effects of physical properties of bagging papers and changes of microclimate in the bags on coloration and quality of peach fruits. J Korean Soc Hortic Sci 44:483–488

127. Sánchez García A, Perera González S, Velázquez Barrera ME et al (2014) Evaluación de la técnica del embolsado sobre la incidencia de la mosca de la fruta (*Ceratitis capitata*) en duraznos. Servicio Técnico de Agricultura y desarrollo Rural. Información Técnica. Agrocabildo, 1–35

128. Bilck AP, Roberto SR, Grossmann MVE et al (2011) Efficacy of some biodegradable films as pre-harvest covering material for guava. Sci Hortic 130:341–343

129. Fan XT, Mattheis JP (1998) Bagging 'Fuji' apples during fruit development affects color development and storage quality. HortScience 33:1235–1238

130. Wang S, Yu L, Li X et al (2003) Effects of cultural techniques on the fruit appearance of bagged Fuji apples. China Fruits 4:36–38

131. Omar AE, El Shemy MA (2014) Enhancing development, rate of ripening and quality of date palm fruit (*Phoenix dactylifera* L.) cv. Zaghloul by bagging pre-harvest treatment. Int J Modern Agric 3:39–45

132. Chen Z, Zhang S, Znang F et al (2003) Ecological effects of bagging on actinidia fruits. Chin J Appl Ecol 14:1829–1832

133. Bugante R, Lizada M, Ramos M (1997) Disease control in Philippine 'Carabao' mango with preharvest bagging and postharvest hot water treatment. Acta Hortic 455:797–804

134. Estrada CG (2004) Effect of fruit bagging on sanitation and pigmentation of six mango cultivars. Acta Hortic 645:195–199
135. Hofman PJ, Smith LG, Joyce DC et al (1997) Bagging of mango (*Mangifera indica* cv. 'Keitt') fruit influences fruit quality and mineral composition. Postharvest Biol Tech 12: 83–91
136. Huang C, Yu B, Teng Y et al (2009) Effects of fruit bagging on coloring and related physiology, and qualities of red Chinese sand pears during fruit maturation. Sci Hortic 121:149–158
137. Johns GG, Scott KJ (1989) delayed harvesting of bananas with sealed covers on bunches.1. Modified atmosphere and microclimate inside sealed covers. Aust J Exp Agric 29:719–726
138. Tyas JA, Hofman PJ, Underhill SJR et al (1998) Fruit canopy position and panicle bagging affects yield and quality of 'Tai So' lychee. Sci Hortic 72:203–213
139. Wang J, Chen H, Zhou Q et al (2003) Effects of bagging on the fruit quality in *Litchi chinensis* fruit and pesticide residues in it. Chin J Appl Ecol 14:710–712
140. Yang WH, Zhu XC, Bu JH et al (2009) Effects of bagging on fruit development and quality in cross-winter off-season longan. Sci Hortic 120:194–200
141. Barreto LF, Cavallari LDL, Venturini GC et al (2015) Propagação de rambutanzeiro (*Nephelium lappaceum* L.) por enxertia. Revi Caatinga 28:176–182
142. Oliveira RP, Scivittaro WB, Vargas JR (2004) Fita plastica e fita degradavel na enxertia de citros. Rev Bras Frutic 26:564–566
143. Khopade R, Jadav RG (2013) Effect of different grafting dates and wrapping materials on success of softwood grafting in custard apple (*Annona squamosa* L.) cv. Local selection. Int J Process PostHarvest Tech 9:806–808
144. Wazarkar SS, Patel HC, Masu MM et al (2009) Effect of grafting dates and grafting materials on soft wood grafting in sapota [*Manilkara achras.* (Mill Fosberg)] under Gujarat agroclimatic conditions. Asian J Hortic 4:434–439
145. Becker H, Hiller MH (1977) Hygiene in modern bench-grafting. Am J Enol Vitic 28: 113–118
146. Fournet J (2003) Use of microcrystalline wax composition, preferably mixture of branched paraffins and saturated cyclic hydrocarbons for protecting graft sites in the over-grafting of vines. Patent number FR2834616-A1
147. Blanke MM (2008) Alternatives to reflective mulch cloth (Extenday™) for apple under hail net? Sci Hortic 116:223–226
148. Coventry JM, Fisher KH, Strommer JN et al (2005) Reflective mulch to enhance berry quality in Ontario wine grapes. Acta Hortic 689:95–101
149. Layne DR, Jiang ZW, Rushing JW (2001) Tree fruit reflective film improves red skin coloration and advances maturity in peach. HortTechnology 11:234–242
150. Kim EJ, Choi DG, Jin SN (2008) Effect of pre-harvest reflective mulch on growth and fruit quality of plum (*Prunus domestica* L.). Acta Hortic 772:323–326
151. Katan J (2014) Three decades of soil solarisation: achievements and limitations. Acta Hortic 1015:69–78
152. Elmore C, Stapelton J, Bell C et al (1997) Soil solarization. A nonpesticidal method for controlling diseases, nematodes, and weeds. University of California Div. Agriculture and Natural Resources. Publication 21377. Oakland, pp 1–17
153. Manera C, Margiotta S, Di Muro E et al (2002) Experimental tests on innovative and biodegradable films for solarisation soil in a site of South Italy. Acta Hortic 578:363–371
154. Scopa A, Candido V, Dumontet S et al (2008) Greenhouse solarisation: effects on soil microbiological parameters and agronomic aspects. Sci Hortic 116:98–103
155. Castronuovo D, Candido V, Margiotta S et al (2005) Potential of a corn starch-based biodegradable plastic film for soil solarization. Acta Hortic 698:201–206
156. Candido V, Miccolis V, Basile M et al (2005) Soil solarization for the control of *Meloidogyne javanica* on eggplant in southern Italy. Acta Hortic 698:195–200
157. Russo G, Candura A, Scarascia-Mugnozza G (2005) Soil solarisation with biodegradable plastic film: two years of experimental tests. Acta Hortic 691:717–724

158. Bonanomi G, Chiurazzi M, Caporaso S et al (2008) Soil solarisation with biodegradable materials and its impact on soil microbial communities. Soil Biol Biochem 40:1989–1998
159. Candido V, D'Addabbo T, Miccolis V et al (2011) Weed control and yield response of soil solarisation with different plastic films in lettuce. Sci Hortic 30:491–497
160. Candido V, D'Addabbo T, Miccolis V et al (2012) Effect of different solarising materials on weed suppression and lettuce response. Phytoparasitica 40:185–194
161. Raj H (2014) Comparative efficacy of biodegradable plastic and low density polyethylene mulch on viability of soilborne plant pathogens of strawberry. Indian Phytopath 67:402–406
162. Goldberger JR, Jones RE, Miles CA et al (2013) Barriers and bridges to the adoption of biodegradable plastic mulches for US specialty crop production. Renew Agric Food Syst 30:143–153

Chapter 5
Biodegradable Spray Mulching and Nursery Pots: New Frontiers for Research

Gabriella Santagata, Evelia Schettini, Giuliano Vox, Barbara Immirzi, Giacomo Scarascia Mugnozza and Mario Malinconico

Abstract Agricultural activities need plastics for many applications such as films for soil mulching and pots for plants transplanting. The use of plastic products, made of fossil raw materials, such as polystyrene, polyethylene, and polypropylene results in huge quantities of plastic wastes to be disposed of. In the past two decades, the growing environmental awareness strongly encouraged researchers and industries toward the use of biodegradable polymers for solving the plastic waste problem. Researchers have made strong efforts to identify new biopolymers coming from renewable sources as valid ecosustainable alternatives to petroleum based plastic commodities. The main research results and current applications concerning the biodegradable plastics in agriculture, such as thermo-extruded Mater-Bi and sprayable water-born polysaccharides based coatings, are described in this chapter. A lineup of biopolymers coming from raw and renewable sources, such as polysaccharides, are reported; the intrinsic chemico-physical properties of polysaccharides, responsible for the realization of dry water stable hydrogels, suitable for the formation of both soil mulching coatings and transplanting biopots, are investigated. A description of the natural additives, fillers and cellulosic fibers included in the polymeric matrices, able to enhance the mechanical performance of coatings and pots is provided, together with the outputs in the specific applications.

Keywords Renewable by-products · Ecosustainability · Biocomposite · Spray technique · Sustainable agriculture · Biodegradable pots · Biodegradable mulches

G. Santagata · B. Immirzi · M. Malinconico (✉)
Institute for Polymers, Composites and Biomaterials (IPCB-CNR),
Via Campi Flegrei 34, 80078 Pozzuoli, NA, Italy
e-mail: mario.malinconico@ipcb.cnr.it

E. Schettini · G. Vox · G. Scarascia Mugnozza
Department of Agricultural and Environmental Science (DISAAT),
University of Bari, Via Amendola 165/a, 70126 Bari, Italy

Fig. 5.1 Straw mulching (**a**) and LDPE black mulching (**b**)

5.1 Introduction

Mulching is a worldwide agricultural practice consisting in covering the soil with a natural or synthetic material in order to provide suitable conditions for plant growth, to conserve moisture, to prevent weed and nutrient leaching, and to provide a barrier to soil pathogens [1–4]. This technique was widely performed in the past by using natural mulches, such as straw, leaves, fibers, and compost (Fig. 5.1a), while over the past decades it has undergone progressive changing in both methods and perspectives by the introduction and application of a new generation of synthetic plastic based materials. In particular, the most widely mulches nowadays used on large scale are plastic films made with low-density polyethylene (LDPE) (Fig. 5.1b) [5–7]. Mulching plastic films have slits or holes through which plants grow. Plastic films, being impermeable to water, reduce the loss of moisture from the soil: when water evaporates, water condenses on the underneath surface of mulching film, dropping on the topsoil. Thus, moisture is assured for a long period inducing an efficient water-saving between two following irrigations. The reduction of water evaporation from the soil avoids both the formation of substrate scabs and the surface soil erosion, preserving the soil structure during the crop period. Finally, as water saving is concerned, synthetic mulching avoids the rising of water containing salts, which is a crucial target in countries where the saline amount in water sources is very high.

5.2 Traditional Mulching: Use, Advantages
and Drawbacks

Traditional mulching films can be characterized by different radiometric properties aimed to satisfy specific requirements; so the plastic films can be black, transparent, photoselective, and reflective [8]. Black mulching films are the most used worldwide mulches because they induce the suppression of spontaneous weeds growing, as they avoid penetration of photosynthetically active radiation (PAR). Transparent mulching films increase soil temperature since solar radiation energy, passing throughout the film, heats the soil beneath; the heat trapped is diffused in the deeper layers of the soil and preserved during the night because of the "greenhouse effect" induced by the film [2]. Using transparent films, seeds can germinate quickly and young plants can rapidly establish a strong root growth system and suitable crop growing conditions but at the same time weed growth can be encouraged. In order to improve the soil heating avoiding the growth of spontaneous weeds, mulching photoselective films can be used: they transmit a high fraction of solar infrared radiation blocking most photosynthetically active radiation that induces weed growth [9].

Reflective mulching films are opportunely colored in relation to peculiar task. Silver, yellow, white, and aluminised films can delay and reduce the incidence of aphid-borne viruses [10]. Reflective red films can promote the yield and the colour of some crops, such as pepper, radish, tomato, and strawberry, enhancing their flavor too [11]; opaque films, white as well as white-on black films, are commonly used in tropical regions, where the soil temperature is too high, to prevent germination of annual weeds by reflecting most of the incident solar radiation (Fig. 5.2) [12, 13].

Plastic mulching films must guarantee their mechanical and physical performances during the whole crop cycles, in relation to the cultivation needs, the geographical region and the season of cultivation; films must be easily handled during the phase of settling on the soil and they should have a life-time long enough to assure an easy removal from the soil, at the end of crop production [14–16]. Oil-derived plastics, such as high-density polyethylene, low-density polyethylene, linear low-density polyethylene, and polypropylene, well match the requested requirements [12]. Moreover, the commercial polyolefin films display good radiometric properties, resistance to microbial attack and to thermo-photo-degradation, easy processability and, last but not least, low cost. All these properties are responsible for the widespread diffusion and consumption of petroleum derived plastic films for mulching. Every year in agriculture at least 1 million tons of plastic mulch film is used worldwide [17]. It is estimated that up to 2020, the

Fig. 5.2 Photoselective mulching films

growth in agricultural plastic demand at global scale will account for an overall average increase of about 10% per year [18]. The increasing demand for high-quality crops in controlled and intensive agricultural productions encourages farmers to consume materials coming from nonrenewable resources, with consequent expenditure of petroleum through plastic manufacturing, and, above all, with following detrimental plastic waste production [19–23]. Nevertheless, while China is the current leader in the Asian market of agricultural mulching films, followed by developing area of India, in Europe and North America a slower growth of oil-derived plastic film market is occurring. Nowadays, the strict regulation for using petroleum plastics in mulching activity is driving the market towards the increasing demand of biodegradable films [24, 25].

The awareness related to the environmental problems due to the wide use of petroleum plastics arises from the marked drawbacks due to the way of films disposal after their lifetime. At the end of cultivation, plastic mulching films are dirty of soil, fertilizer, and biological wastes, as well as pesticide. These contaminants can reach up to 40–50% by weight; since plastic films with more than 5% contaminants by weight are not accepted for recycling, their recovering and cleaning is too time and hand labor consuming, which is conveyed in unsustainable costs for the farmers [17, 26].

In addition, during exposure in the field, the plastic films undergo photo-degradation process, inducing a reduction of film performances but not its permanence on the soil [27–31]. Since regular gathering, discarding and recycling processes of films are much expensive, plastics are often discarded in common dump or on the side of the street or, even worse, burned with the subsequent emission of toxic substances both in the atmosphere and into the soil [32]. The long permanence of mulching film in the environment causes a huge and unmanageable accumulation, seriously harmful for the environment and for human health [33].

5.2.1 New Eco-friendly Biodegradable Thermoplastic Films

In order to overcome the harmful environmental impact of petroleum-based plastics, scientific research has been focusing the attention on biodegradable materials based on polymers coming from renewable sources (Fig. 5.3a) [34, 35], as valid alternative to oil-derived polymers in packaging and agricultural applications. When disposed in bioactive environments, biodegradable polymers can be degraded by the enzymatic action of microorganisms—bacteria, fungi, and algae—and converted into biomass, carbon dioxide, water, or methane depending on if the degradative environment is aerobic or anaerobic. Hence, at the end of their lifetime biodegradable films may be left on the soil or buried in it and biodegradation will start by means of bacteria flora or they may be blended with other organic material in order to generate carbon rich compost (Fig. 5.3b) [8, 36, 37].

Fig. 5.3 Biodegradable thermoplastic films: **a** during crop cultivation, **b** after cultivation and milling

Most of biodegradable mulches commercially available are starch-based films prepared using thermoplastic processing technology. In order to improve the poor mechanical properties of starch, blends with other polymers and/or plasticizers are developed. Some of the products currently on the market that contain starch are Biosafe™ (Xinfu Pharmaceutical Co., China), Eco-Flex® (BASF, Germany), Ingeo (NatureWorks, USA) and Mater-Bi (Novamont, Italy) [38].

In addition to starch, thermoplastic polymers such as polylactic acid (PLA) and polyhydroxyalkanoates (PHAs) can be used for future perspectives in thermoplastic mulching films. PLA is highly versatile, biodegradable polyester derived from 100% renewable resources, such as corn and sugar beet starch, by means of microorganism conversion of starch into lactic acid molecules through fermentation, and their following organization in macromolecular chains. PLA is a relatively inexpensive biopolymer to manufacture ($\sim$$0.95 per lb), and can be produced in large quantities [39, 40]. PHAs or "green" polymers are promising biodegradable materials obtained by the bacterial fermentation of sugars and/or lipids, even if the primary PHAs sources are bacteria [41]. New experimental agricultural mulches have been prepared from PLA and PHA blends using nonwovens textile technology [38, 42]. Even if the above discussed thermoplastic polymers come from renewable sources, their process involves the employment of additives, plasticizers, and/or lubricants whose environmental impact may be a major concern in organic as well as conventional crop production. Some additives are chemically processed and are considered synthetic material by National Organic Program (NOP) standards [43], so they are avoided in some organic agriculture. Moreover, the NOP standards consider PLA as a synthetic polymer because it is chemically polymerized [44].

Finally, many mulches claiming to be "biodegradable", i.e., matching the ASTM WK29802 [45], are actually "compostable", i.e., able to fulfill the requirements of ASTM D6400 [46, 47].

5.3 Novel Generation of Mulches with Amazing Perspectives: Sprayable Water Solutions, Based on Polysaccharide Formulations

5.3.1 Introduction

An innovative approach of mulch forming is the use of the spray methodology, often used for agricultural application of fertilizers, pesticides, and substances useful for the plant health [8, 48, 49]. In particular, water solutions based on natural polymers, as polysaccharides, are sprayed on the soil in order to form a protective mulching geomembrane, after water solvent evaporation. Polysaccharides represent very interesting natural sources as matrices of sprayable water solutions, due to their abundance, easy availability and fast renewability [50]. They strongly interact with water molecules, forming three-dimensional network, hydrogels, able to swell and retain a significant amount of water within their structure, without dissolving in water. By definition, water must represent at least 10% of the total weight (or volume), being able to reach up to 95% of the total weight (or volume), as in the case of super absorbent materials [51–53] (Fig. 5.4).

The ability of polysaccharides to absorb water is due to hydrophilic functional groups attached to the polymeric backbone, while their resistance to dissolution is owed to chemical or physical cross-links between macromolecular chains. As a matter of fact, in order to be sprayed on the soil, polysaccharides must be water soluble; nevertheless, upon water solvent evolution, the mulching coatings formed need to become water resistant. Indeed, the coating formed on the top soil must assure its permanence and its covering function for all the cultivation time. Actually, exploiting some intrinsic chemical–physical properties of polysaccharides, it is possible to obtain dry hydrogels in form of water stable coatings. This outcome is due to some structural polymeric chain reorganization, such as retrogradation, gel formation process, pH change and ionic cross linking, leading to the development of packed, three-dimensional networks [50].

Fig. 5.4 Dry and swollen polysaccharide based hydrogel

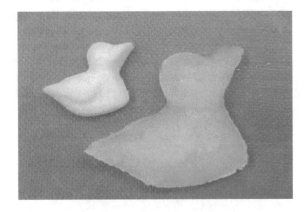

Sodium alginate, galactomannans (guar gum, locust bean gum), agarose and chitosan have been selected and investigated among polysaccharides able to form dry coatings [8, 54–56].

The use of spray coatings does not involve any particular change in usual agronomical practices. Indeed, the new polymeric water-born solutions can be applied by means of airbrushes commonly employed by farmers to spread fertilizers, hormones, and other chemicals useful for plants' health. In addition, soil irrigation practice, commonly occurring by means of drip, hose, and porous tube is not invalidated by the presence of spray mulches. A further advantage of spray technique may concern the avoiding of film layer machines employment, necessary to the setting up and removal of the plastic films. In the water-born formulations, fillers, such as cellulose fibers, carbon black, fine bran of wheat and powdered seaweeds, can be added to the polymeric matrices both to improve the mulching function and to enhance the tensile strength of the coating formed upon drying. Moreover plasticizing polymers, such as hydroxyethylcellulose and natural plasticizers, such as glycerol and polyglycerol, are included in the aqueous polymeric blends to improve the mechanical durability of the soil mulching coatings. Finally, before spraying, the side slope of raised beds should be limited in order to avoid a possible sliding of the water-born coating at the liquid state during the spraying (Fig. 5.5a, b). In pot cultivation, the level of the growing media must be lower than the edge of the pot in order to contain the spray coatings. In presence of plants a protection must be used to maintain stem and leafs clean during the spraying (Fig. 5.5c). In case of plants transplanting, holes can be performed when coating drying process is completed (Fig. 5.6a).

The spray coatings have been initially developed and tested within the Project "Biodegradable coverages for sustainable agriculture BIO.CO.AGRI." (BIO.CO. AGRI, 2001–2005), funded by the European Commission [8]. Anyway, in other following projects targeted on agricultural topics, several experimental trials have been performed in order to follow, improve, and assess different parameters concerning the spray mulches properties, corroborating the effectiveness of the selected polysaccharide based formulations or modulating them on the base of the specific agronomic experimental requests.

Some of the natural polymers are undergoing the previous macromolecular chains rearrangements and therefore selected in the frame of BIO.CO.AGRI. project

Fig. 5.5 Spray mulching coatings: **a** white sodium alginate; **b** black sodium alginate; **c** spray mulching application on pots

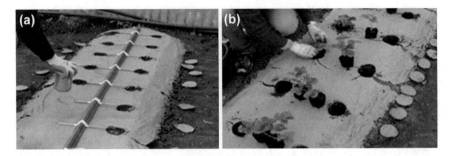

Fig. 5.6 Manual piercing of the coating (**a**); strawberry plant sowing (**b**)

come both from terrestrial origin, such as Arabic Gums [57] and from marine origin, such as Agarose [58], Sodium Alginate [54], and chitosan, this was last mostly derived from wastes of crustacean shells [59, 60]. Water solution of polysaccharides were added with natural plasticizers, such as glycerol and polyglycerol, in order to enhance the mechanical elasticity of the coating and with cellulose fibers, coming from wastes of agro food industry, intended to strengthen the coating texture on the soil. Moreover, colored fillers, such as carbon black, bran of wheat and powdered seaweeds, were included inside the water-borne formulation in order to improve the covering action of the geomembrane (Fig. 5.5).

5.3.2 Sodium Alginate Based Spray Solution

The first polysaccharide tested in the frame of BIO.CO.AGRI. project was sodium alginate, the sodium salt of alginic acid, the structural component of intercellular walls of brown seaweeds, *Phaeophyceae*. Alginic acid confers both strength and flexibility to the algal tissue. It exists in the form of insoluble gel of mixed calcium, magnesium, sodium, and potassium salts, and it is extracted from the grounded thallium upon the collapse and subsequent transformation of tissue in a brown mass [61] (Fig. 5.7a). Alginates are linear water-soluble polysaccharides formed by

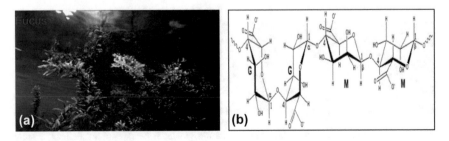

Fig. 5.7 Phaeophyceae brown seaweeds (**a**); alginate chemical composition (**b**)

polymannuronic acid (MM), polygluronic acid (GG) and a mixed polymer (MG), where sequences like GGM and MMG coexist too (Fig. 5.7b) [62, 63].

The mannuronic acid forms β (1–4) linkage, so that M-block segments show linear and flexible conformation; the guluronic acid, differently, gives rise to α (1–4) linkage, introducing in this way a steric hindrance around the carboxyl groups; for this reason the G-block segments provides folded and rigid structural conformations, responsible of a pronounced stiffness of the molecular chains [62, 64, 65]. The great interest with sodium alginate is strictly related to its peculiar gelling properties; in fact, alginate solutions can crosslink in presence of divalent ions which cooperatively interact with blocks of guluronic units to form ionic bridges between different chains (Fig. 5.8a) [66, 67]. The most popular model to account for the chain-to-chain association is the "egg box model" (Fig. 5.8b) [68, 69]. In this model, two carboxylated groups of adjacent α-L-guluronate residues of GG homopolymeric blocks interact with calcium ions, physically embodying them in cavities similar to cardboard egg box. In this way, a stable, continuous and thermo-irreversible three-dimensional network forms [65, 68]. According to some authors [70, 71], it seems that although calcium ions is localized in egg-box arrangements, macromolecular chains still promote lateral association [72].

Since the strong external gelation occurring between sodium alginate and available calcium ions, this polysaccharide was chosen as matrix of mulching-sprayable water solutions [73].

Indeed, once sprayed, the polysaccharide suddenly interacts with calcium ions naturally present in the soil, forming a strong water stable network, whose durability is compatible with the agricultural cultivation lifetime (Fig. 5.4).

To improve the elasticity of the sodium alginate (A), Hydroxyethylcellulose (HeCell) was introduced in the water solutions. Hydroxyethylcellulose (HeCell) is a biodegradable non-ionic water-soluble cellulose ether, obtained by the introduction of hydroxylethyl groups in the repetitive chains of glucosidic units (Fig. 5.9a). It forms homogeneous blends with sodium alginate [74], and semi-interpenetrating network (SIPN) with calcium alginate, in this way assuring its permanence in crosslinked polysaccharide geomembrane during all the cultivation time [75]. In polymeric based solution, polyglycerol was introduced as plasticizer in order to

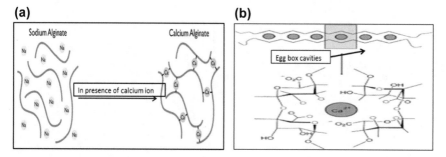

Fig. 5.8 Calcium ionic bridges in alginate hydrogel (**a**); egg-box model for binding of divalent cations to α-L-guluronate residues and binding sites in GG sequence (**b**)

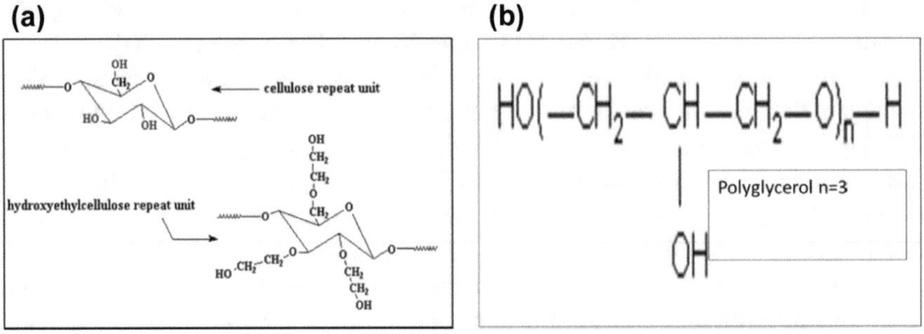

(a)   **(b)**

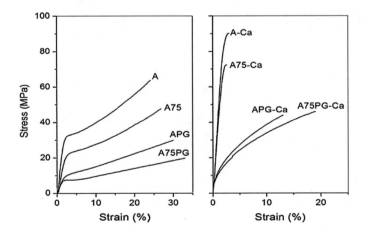

Fig. 5.9 Hydroxyethylcellulose macromolecular structure (**a**); polyglycerol structure (**b**)

improve the mechanical response of the coating on the soil. Polyglycerol (PG) consists of glycerol molecules bonded by an ether linkage (Fig. 5.9b). The presence of larger amount of hydroxyl groups enhances its interaction with polar groups of both alginate and HeCell, by means of hydrogen bonding. This outcome, confirmed by laboratory tests performed on precasted coatings [64], could induce a prolonged permanence of the plasticizer in the coatings. Table 5.1 shows the different compositions of alginate based mulches (MA) while Fig. 5.10 evidences their stress–strain curves before and after external gelation process in calcium chloride. From the data, it is meaningful to observe that the addition of HeCell and PG to A, as expected, causes a progressive decrease of both Young's modulus and stress at break level during the deformation, to the benefit of strain at break, particularly increasing in A75-PG formulation, for this reason chosen for the experimental test.

Moreover, it is worthy of consideration that after cross-linking process, all dried and conditioned samples evidence a steady increasing of thickness (Th), as shown in Table 5.2.

Fig. 5.10 Stress-strain curves of alginate based films

Table 5.1 Sample composition of MA

Sample	A (g)	HeCell (g)	PG (g)
A	3.00	–	–
A-PG	3.00	–	1.00
A75	2.25	0.75	–
A75-PG	2.25	0.75	1.00
HeCell	–	3.00	–

Table 5.2 Thickness of uncross-linked (Thuncr.) and cross-linked samples (Thcr) and percentage increase (ΔTh%)

Sample	Thuncr. samples (μm)	Thcr samples (μm)	ΔTh increase (%)
A	74	107	45
A-PG	80	110	38
A75	75	118	57
A75-PG	81	125	54

This outcome, previously detailed by Russo et al. [64–72], underlines that the cross-linking points, introduced during external gelation of films in water solution of calcium chloride, induce the samples swelling and fix their increased thicknesses even in dried samples. This outstanding result will be exploited to investigate the water saving provided by different spray mulches with respect to commercial plastic films. Indeed, as mentioned above, hydrogels, being three-dimensional cross-linked hydrophilic polymer networks are able to swell or de-swell reversibly in water, retaining large volume of liquid in the swollen state. Hence, they can be designed or formulated with controllable responses, as to shrink or expand, consequently to soil cultivation requirements.

MA mulches were firstly tested within the Project "Biodegradable coverages for sustainable agriculture BIO.CO.AGRI." (BIO.CO.AGRI, 2001–2005), funded by the European Commission. The water polysaccharide-based solution was sprayed on the soil during strawberry cultivation performed in soil [8]. Onto the soil a powder mixture of seaweeds flour and fine bran of wheat was distributed uniformly to provide a fibrous bed (Fig. 5.11a) and the polysaccharide-based water solution was applied by means of an airbrush, using a high pressure spray machine, commonly used for agricultural practices. The volume sprayed accounts for the coating's thickness and its lifetime, tailored on the specific cultivation (Fig. 5.11b). After at least 24 h, time requested for solvent evaporation, the coating formed and the seedling transplanting could be done, by holing the coating in correspondence of plant sowing (Fig. 5.6).

In terms of functionality, durability, and agronomic response, it is not feasible to standardize a model support, since the effects of spray mulching are strictly linked to the real interaction between the sprayed coating and the soil. As an example, it is not possible to follow the mechanical response of the composites with the time by

Fig. 5.11 Powder mixing deposition (**a**); MA solution spraying (**b**)

Fig. 5.12 Spherical dart (**a**), sample broken by dart penetration (**b**), curve of displacement vs time (**c**)

using the standard tensile tests approached for plastic mulching films, as the spray coatings are not self-standing materials; their consistency is supported by the soil beneath and it is object of several fluctuant variables. Nevertheless, it was possible to test their mechanical behavior by means of an empiric test called "puncture tests", widely detailed by Malinconico et al. [8]. Briefly, this method consists in penetrating the samples, opportunely fixed on a metal support, with a dart moving under the action of a compression force (Fig. 5.12a, b). Recording the displacement of the dart inside the sample up to sample rupture as a function of the time, it was possible to monitor the increasing of composite stiffness (decreasing of displacement), likely due to the loosing of plasticizer (Fig. 5.12c). The specimens tested were obtained by cutting the mulches.

Besides sodium alginate, other polysaccharides, such as galactomannans, have been tested as new water-borne sprayable mulches.

5.3.3 Galactomannan–Agar-Based Spray Solution

Galactomannans are polysaccharides extracted by seeds of leguminous plants (Guar Gum) and Carob tree (Locust bean gum). They are heterogeneous polysaccharides formed by main chain made of (1–4)-linked D-mannopyranose (Man) units, to which (1–6) linked D-galactopyranose (Gal) residues are attached (Fig. 5.13a). Variations

(a)   **(b)**

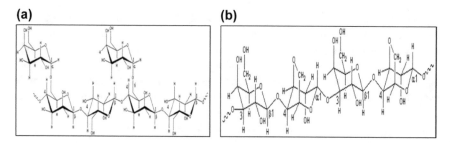

Fig. 5.13 Chemical structure of galactomannans (**a**) and agar (**b**)

in Man/Gal ratio cause significant changes in solubility, viscosity and ability to form a gel. Generally, polysaccharides with higher Gal content, as guar gum, easily dissolve in water due to the presence of more side chains, even if their tendency to form a gel is very low. In contrast, carob gums show higher Man content, consisting of long unsubstituted macromolecular chains with few lateral residues, able both to provide self-aggregations and to interact with other gelling polysaccharides in order to provide water insoluble gels [76]. As a matter of fact, in aqueous solutions, galactomannans exist in a random coil conformation as shown in Fig. 5.14a; nevertheless more-ordered forms, following the macromolecular self assembling (Fig. 5.14b) (retrogradation) can occur upon solvent evaporation, providing more packed structures where strong hydrogen bonds develop (Fig. 5.14c). Such a structure, characterized by the macromolecular re-association in ordered structure (crystalline aggregates), provides the formation of water resistant coatings [77].

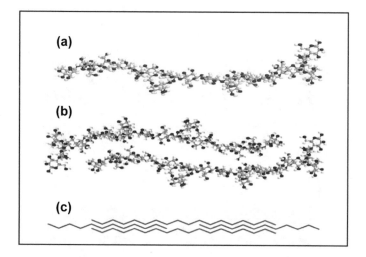

Fig. 5.14 Galactomannans ribbon conformation (**a**), self-aggregation (**b**) and retrogradation process (**c**)

Nevertheless, in order to strengthen the mechanical performance of the covering coatings on the soil after water removal, small amounts of agarose were added. Agar is a gel-forming polysaccharide extracted from *Rhodophyceae* seaweeds with a sugar skeleton consisting of alternating 1,3-linked β-D-galactopyranose and 1,4-linked 3,6 anhydro-α-L-galactopyranose units (Fig. 5.13b); it is used as a model biopolymer in gelation [78]. Agar contains mainly agarose but also agaropectin; the former contributes to gelation while the latter weakens the gel. Agar can be solubilized at around 70–80 °C and in solution the polymers take the form of random coils. Upon cooling the three equatorial hydrogen atoms on the 3,6-anhydro-L-galactose residues constrain the molecule to form a right-handed double helix, that linking each other, form bundles of right-handed double helices (Fig. 5.15a). At junction zones the bundles interact, thus forming a three-dimensional network able to immobilize water molecules in its interstices [79]. In this way, thermo-reversible gels form. Regarding its gelling power, agar is outstanding among other hydrocolloids. Agar gels can be formed in very dilute solutions, containing a fraction of 0.5–1.0% of polysaccharide. These gels clearly demonstrate the interesting phenomenon of both syneresis, spontaneous extrusion of water through the surface of the gel, and hysteresis, temperature interval between melting and gelling temperatures. The hysteresis range is wide enough to assure the formation of gel between 32 and 45 °C and its melting at 85 °C.

As a matter of fact, by means of synergic physical interaction between galactomannans and agar, blended in water solution in suitable proportion, it is possible to obtain a double reinforced gel at experimental conditions far below those required for gelation of the single polymers. In the specific case a 2% (w/v) water solution of Guar gum and Locust bean gum in the proportion 75:25 (w/w) and a non-gelling concentration of agarose 0.05% (w/v) was dissolved at room

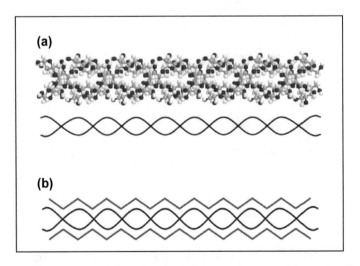

Fig. 5.15 Double helix agar gel formation (**a**), Galactomannans-agar gel interaction (**b**)

temperature. The low concentration of agar was not necessary to raise the temperature up in order to dissolve the polysaccharide. Such a solution was easily sprayed on the soil due to the non-gelling concentration of agar. As a consequence, upon water removal, a solid, clear, and water proof coating was obtained, that in wet conditions could swell to a gel without dissolving. As a matter of fact, during the concentration phase on the soil, besides retrogradation process involving only galactomannan macromolecular chains (Fig. 5.14b), strong physical associations between galactomannan ribbon conformations and native double-helical stretches of the algal polysaccharide chains occurred; so it is plausible that the polymeric matrix residues perfectly fits a double helix polysaccharide conformation to form a packing pattern (Fig. 5.15b). Finally this complex three-dimensional network can physically entrap retrograded ribbon galactomannan chains, providing a strong water stable gel (Fig. 5.16).

To assure the mulching power and to improve both the mechanical response and soil permanence of the mulching coatings, cellulose fibers and natural plasticizers such as glycerol and polyglycerol were included in the aqueous polymeric blends. Moreover, with the aim to enhance the mulching power, carbon black was also added to the formulation.

This polysaccharide-based system (identification code MGA) was tested on lettuce cultivation both in open field and in greenhouse (Fig. 5.17a, b) in the experimental fields of the University of Applied Science of Osnabruck, in order to follow the spray trials in more severe climatic conditions. As an example, the open experimental field sprayed with MGA water solution has been reported in Fig. 5.18a, whereas a particular region of the treated soil has been evidenced both at initial time (Fig. 5.18b) and after 60 days (Fig. 5.18c) of coating permanence on

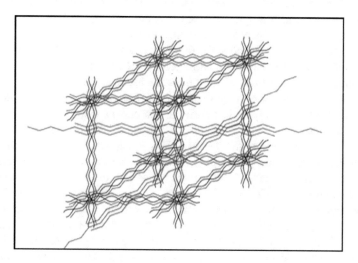

Fig. 5.16 Three dimensional gel network entrapping retrograded galactomannans in a double reinforced gel

Fig. 5.17 Experimental setting of galactomannan spray mulching: open field (**a**) and greenhouse (**b**)

Fig. 5.18 MGA spray formulation on lettuce (**a**), zoom of sprayed soil $t = t0$ (**b**), zoom of sprayed soil $t = 60$ days (**c**)

the soil. It is worthy to observe that the natural aging of the geomembrane provides a drastic reduction of the polysaccharide amorphous fraction, easily metabolized by bacterial flora of the soil. Nevertheless, the crystalline cellulose fraction is still discernible. This macroscopic evidence is confirmed by surface morphological analysis performed by means of scanning electron microscopy (Fig. 5.19), on samples picked up at $t = t0$ and at $t = 60$ days. It is worthy to observe that the neat polymers show a homogeneous distribution of cellulose fibers between the polymeric matrix and a deep embedding of the same inside the macromolecular network (Fig. 5.19a). After 60 days of permanence on the soil, the fibers are still well evidenced, while the polymeric matrix is not clearly visible (Fig. 5.19b). This outcome suggests that, after 60 days, the polymeric amorphous fraction of the composite, more sensible to the bacterial flora attack, starts to undergo the degradation and biodegradation processes, whereas the cellulose fibers, characterized by crystalline ordered structure, are more resistant to the disintegration and biodegradation action of microorganisms. Apart from the interesting performance of these natural gums, it is also worthy to underscore that the raw materials (locust bean and guar) from which such polysaccharides are extracted are much less expensive than e.g., agar; so mulching gels, coatings and geomembranes of specified and enhanced properties can be obtained by mixing inexpensive

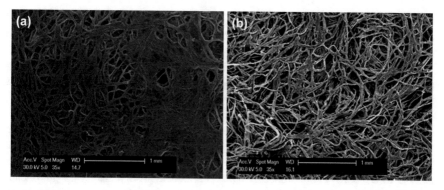

Fig. 5.19 SEM micrograph of MGA sample at time $t = t0$ (**a**) and at time $t = 60$ days (**b**)

galactomannans with small amount of agarose, thus obtaining new cost-effective and ecosustainable polysaccharide-based systems.

5.3.4 Chitosan Based Spray Solution

Another biopolymer widely used and experimented in the frame of new water-borne sprayable mulching formulations is chitosan, a linear polysaccharide composed of (1-4)-2-acetamido-2-deoxy-ß-D-glucan (*N*-acetyl D-glucosamine) and (1-4)-2-amino-2-deoxy-ß-D-glucan (D-glucosamine) units (Fig. 5.20). As such, chitosan is not extensively present in the environment; however, it can be easily derived from the partial alkaline deacetylation of chitin, a homopolymer of 1-4 linked 2-acetamido-2-deoxy-ß-D-glucopyranose. Chitin is the major structural support of

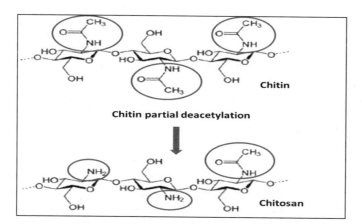

Fig. 5.20 Chitin and chitosan chemical structure

crustaceans and insects exoskeleton and the cell walls of fungi and some algae [80] and, in terms of availability, the third most abundant polysaccharide occurring in nature after cellulose and starch [8, 81].

Most of chitosan physicochemical functionalities are due to the presence of both amino groups at the C-2 position and hydroxyl groups at C-6 and C-3 position of the glucose residues; due to the easy protonation of the amine groups at acidic pH, chitosan is considered a cationic polyelectrolyte, with a positive charge extent strictly linked to its molecular weight, deacetylation degree, chitin sources and ultimately to the method of extraction process. Differently from chitin, which is insoluble in common solvent for less than chemical modifications, chitosan is soluble in aqueous acidic solutions where the amine groups are easily protonated [82]. The wide and versatile range of chitosan functions and applications is therefore related to its polycationic behavior, which is unique among polysaccharides and natural polymers [81]; thus chitosan is commonly used like antimicrobial, antioxidant and wound healing accelerating agent in pharmaceutical and medical industries [60, 83] as well as in food, biotechnological, agricultural, and environmental applications [84, 85].

Moreover, as a result of intra- and intermolecular hydrogen bonding, high molecular weight chitosan shows good coating-forming properties [86]; nevertheless, the industrial employment of chitosan-based coatings is limited due to their poor physical–chemical and mechanical performances. Therefore, research has been paying attention to biodegradable chitosan-based systems, such as blends and biocomposites, in which the polysaccharide is blended with other polymers and/or mixed with reinforcing fibers, plasticizers, and additives [87]. Siró and Plackett [88] evidenced an improvement of chitosan mechanical performances by blending it with microfibrillated cellulose; in this study the authors showed a suitable physical interaction at the interfacial region between the two polymers. This outcome, confirmed by other authors, is in all likelihood due to the similar primary chemical structure of chitosan and cellulose: the two polymers show the same β-glycosidic linkage and differ only for the different polar groups present on the C-2 binding sites: primary amino groups in chitosan (Fig. 5.20) and hydroxyl groups in cellulose (Fig. 5.9a) [89]. Thus, combining the physicochemical properties of chitosan with the excellent mechanical performance of cellulose fibers, it is possible to obtain chitosan–cellulose composite-based coatings characterized by high strength, good biocompatibility, biodegradability, and hydrophilicity.

Hence, chitosan-cellulose based systems (Ch-Cell) were prepared by dissolving chitosan (Ch) in 3% v/v of aqueous solution of acetic acid, adding polyglycerol as plasticizer and including cellulose microfibers (Cell) in order to enhance mechanical resistance of coating on the soil [90].

In order to simulate the natural weathering endured by chitosan-cellulose-based coatings during their permanence on the soil, the biocomposites were exposed to accelerated photo-degradation process by means of ultraviolet radiation and moisture, in the form of condensation technique. The photo-degraded samples were periodically recovered (after 0, 100, 200, 300, 400 h of UV exposure) and together with the un-aged coatings were investigated by means of thermal properties and morphological analysis. When dealing with polymer based blends, the miscibility between the constituents is an important factor in the development of new materials with enhanced performances with respect to the single polymers. A valid approach to estimate the miscibility of polymers, as widely confirmed by literature data, is the evaluation of the glass transition phenomenon of typical amorphous and semi-crystalline polymers. It consists of the changes from the glassy state into either a liquid or a rubbery state of a polymer. Although this transition is a gradual one, and the glass transition phenomenon spans a wide temperature window, experimentalists tend to report and tabulate unique values of the so-called glass transition temperature (Tg) [91]. Tg values are particularly needed in case of binary polymer blends; they tell us whether the blends are miscible, semimiscible (called compatible) or not miscible at all. It is generally accepted that the presence of two separate Tg's in polymeric blends provides a strong signature of immiscibility, whereas a single signal of Tg in-between the Tg's of the pure polymers is plainly indicative of molecular miscibility. In Table 5.3 the Tg values of the samples (Ch-Cell) as a function of the photo-aging time are detailed. In all the samples, Tg was taken as the mid-point of the heat capacity step-change in the glass transition measurement.

Experimental data evidence the presence of one Tg value, probably due to the strong physical interaction, via hydrogen bonding, between the cellulose fibers and chitosan [92]. More specifically, the cellulose fibers, interposing between the macromolecular network of polymeric matrix, disturbed the regular packing and the structural organization of polymeric chains. In particular hydrogen bonding interactions between hydroxyl groups of cellulose fibers and hydroxyl and free amine groups of chitosan occurred, providing the increasing of backbone rigidity and a following reduction of macromolecular free volume. This is not surprising considering that, reducing the number of thermally activated chains and their mobility, the glass transition temperature rises up, furthermore diminishing the change in specific heat capacity. In Ch-Cell composites, an increase of Tg values was observed for the first 300 h of UV/Cond. exposure time. Nevertheless, after 400 h of UV photo-aging treatment, a Tg dropping was observed. This result is probably due to the beginning of degradation process provoking a depolymerization of the macromolecular network [93, 94].

Scanning electron microscopy (SEM) performed on the Ch-Cell Samples showed that the polymer underwent to strong surface changes during exposure to

Table 5.3 Tg values of ChCell-based blends as a function of UV aging time	Samples	Tg (°C)
	ChCell0	130.58
	ChCell100	131.22
	ChCell200	133.24
	ChCell300	137.81
	ChCell400	122.76

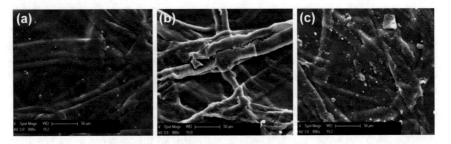

Fig. 5.21 SEM micrographs of Ch-Cell0 (**a**), Ch-Cell300 (**b**), Ch-Cell400 (**c**)

UV-condensation (Fig. 5.21). In particular the untreated coating (Fig. 5.21a) showed that flat and long cellulose fibers form a uniform and random texture coated with a thin and regular layer of polymer. Micrographs of aged samples up to 300 h of UV-condensation exposure evidenced a gradual degradation of both chitosan and fiber (Fig. 5.21b). As a consequence, the chitosan coating started breaking, inducing the removal of the chitosan layer fragments from the exposed surface, whereas the fibers underwent a degradative process causing internal cracks. Finally, the micrographs related to Ch-Cell400 sample (Fig. 5.21c) evidenced a strong degraded surface; the fibers were separated into small short-length fragments while some prints of fibers pulled out from surface could be evidenced too.

5.4   Biodegradable Pots: New Challenge for Eco-friendly Transplanting Practice

In horticulture, a worldwide cultural practice is transplanting, i.e., the process of removing a plant from the place where it has been growing to another growing location that can be soil or a larger container. Some crops do not grow by sowing the seeds directly in the soil, so seeds, bulbs and young plants are allocated at nurseries and greenhouses in cell trays or pots containing a growing substrate in order to grow under uniform and suitable cultivation and microclimate conditions until the transplant occurs. The advantages of the transplanting technique if compared to direct seeding in soil are: higher crop density and uniformity, growth of the seedlings without competition from weeds, prevention of soil pests and diseases, selection for transplanting of only healthy seedlings to improve the growth of vigorous seedlings and plants, facilitation of the use of expensive hybrid seeds and of a wider range of substance suitable for the plant health, and higher yields. Growers use pots and cell trays of different materials, sizes, shapes, and colors to suit crop species, growing methods, and marketing strategies [95, 96]. The pots, made of fossil raw materials, such as polystyrene, polyethylene, and polypropylene are un-permeable and rigid containers so the roots tend to circle the outer perimeter of the root ball, which can result in reduced plant growth, health, and survival once transplanted [96, 97]. Transplanting is necessary in order to promote better development of shoot system of plants, allowing a more natural development of root structure. Nevertheless, during the transplanting, the roots can be damaged; the period immediately after transplanting is a vulnerable one for all the species, so any other stress must be avoided. Most of the transplanting pots are made of petroleum derived materials, such as polypropylene and polyethylene. The use of nonrenewable oil-based plastics is widespread due to their easy processability, their good mechanical properties, long lifetime, high resistance to microbial degradation, and their relatively low cost. After use, plastic pots, contaminated with soil, organic matter, and agrochemicals require a correct collection, disposal, and recycling process that are costly; soften they are neglected in landfill or burnt in an uncontrolled way with the subsequent emission of toxic substances both into the atmosphere and into the soil. Only in Italy each year there are tens thousands of tons of dumped pots [98].

A valid alternative to the employment of petroleum based thermoplastic pots may be represented by the use of biodegradable pots [95, 96, 99–105]. Biodegradable pots must be engineered in such a way that water, air, and roots will easily penetrate the walls of the pot both assuring a healthy roots growing, and cooperating with bacterial flora in the biodegradation process. Biodegradable pots can be planted together with seedlings or young plants directly into soil, guaranteeing fast plant transplanting, field clean up, and no pot disposal, and reducing farm labor, cost, and environmental pollution. Biodegradable pots will allow a more natural development of the roots in the growing substrate both into the soil in open field and into the growing media in larger containers, for example in greenhouse cultivations, avoiding the problems of roots spiraling and binding. The biodegradable pots are subjected to biodegradation process: once buried into the soil, they will be transformed in biomass and inorganic products (e.g., carbon dioxide and water).

Several companies, such as William Sinclair Horticulture Ltd. (Lincoln, England, http://www.william-sinclair.co.uk/), Enviroarc (Scoresby, Australia, http://www.enviroarc.net/), Fertil SA (Boulogne Billancourt, France; www.fertilpot.com), CowPots (East Canaan, Connecticut, United States; http://www.cowpots.com/), Ecoforms (Sebastopol, California, United States; http://ecoforms.com/), Jiffy Products International AS (Stange, Norway, http://www.jiffygroup.com/) etc., produce biodegradable pots made of plant fiber, wood fiber, rice, rice hulls, starch, coir, peat, grasses and vegetable oils, cow manure, etc. The lifetime of these pots ranges from few months to five years depending if the pots are used outdoors or indoors.

Some biodegradable pots need a composting site to completely decompose [106], other ones are characterized by unsuitable mechanical properties hindering the roots to pass throughout, other more may emit odd smell and, last but not least, many of the biodegradable pots are more expensive if compared to traditional plastic pots.

Recently researchers have developed novel biodegradable and cost-competitive pots made as biocomposites whose continuous phase is characterized by biopolymers coming from renewable and available origin [36, 54, 102, 107] and whose solid phase, dispersed within the polymeric matrix, is represented by natural fillers and fibers coming from wastes of agro food and textile processing industries [108]. The arrangement obtained from the phase combination produces a system characterized by improved structural, mechanical, and chemical–physical properties [109, 110].

Among biopolymers, polysaccharides coming from marine origin, such as chitosan and sodium alginate, can be used as binders for pots applications because they are biodegradable, biocompatible, and nontoxic polymers widely available and renewable [36, 54, 111]. Polysaccharides show high affinity with water molecules due to the presence of polar groups (OH^-, NH_3^+, COO^-) on macromolecular chains. Polysaccharides induce the development of hydrogels, i.e., three-dimensional water stable networks, structured by means of ionic, covalent, thermo-reversible or pH-reversible cross-linked processes [72, 112].

Chitosan can be applied due to its polycationic properties, which are unique among abundant polysaccharides and natural polymer in general. Chitosan behaves both as a fungicide, preventing microbial infections when sprayed on plants, and as a fertilizer, accelerating the germination and growth of the plants [113]. It is insoluble in water and this feature is important for its applications in the soil, resisting to the common agricultural practices as plants watering and assuring the binding function of the pots fibers for the life cycle time. Chitosan readily dissolves in dilute solutions of most organic acids such as acetic, citric, tartaric acids. The acetic acid, commonly used in agriculture as an inhibitor of spontaneous weeds growing will be chosen like solvent of chitosan.

Sodium alginate is a water-soluble polysaccharide coming from brown seaweeds belonging to *Phaeophyceae* Family. Due to its polymeric structure, in presence of divalent cations, such as calcium, it gives rise to three-dimensional, thermo-irreversible, stable, and insoluble network (gel) [76, 102] (Fig. 5.8).

Natural fibers, dispersed within the polymeric matrix, act as a reinforcement that is able to enhance the strength and stiffness of the resulting composite structures. The mechanical properties of natural fibers, depending on the plant source, plant age, separating technique, moisture content, etc., are poorer if compared to those of the most widely used competing reinforcing man-made fibers, such as glass, carbon, aramid, etc. [114].

The specific properties (property-to-density ratio), such as strength and stiffness, of plant fibers are comparable with the ones of glass fibers due to the low-density of natural fibers [115]. The presence of plant fibers influences the mechanical properties of biodegradable polymers as it has been widely explored in literature [116, 117]. Polysaccharides and natural fibers, consisting mainly of cellulose, are characterized by chemical similarity and by a highly hydrophilic character of both components thus resulting in an increased tensile strength of the reinforced polymers [109].

Innovative biodegradable pots can be made with sodium alginate, as polymeric matrix, and tomato and hemp fibers, as natural reinforcing dispersed phase [102]. The mechanical properties of the biocomposites are influenced by the physical-chemical interactions among polymeric matrix and fibers [118]. As a matter of fact, the hydrogen bonding between the active functional groups of the biocomposites components could entail suitable effects [119].

The use of fibers coming from wastes of processing industries of tomatoes, citrus fruits, hemp, olives, Artemisia, as well as cellulosic fibers coming from kenaf plants, could have a positive and remarkable impact on the crucial item related to the management of agro food industry wastes [102].

The market costs of sodium alginate and chitosan are elevated due to the high level of purification of polymers requested for their employment in biomedical, pharmaceutical and food industries. In order to make biodegradable pots, it is not necessary to work with purified polymers; on the contrary the presence of protein and filler together with the polysaccharide fraction both represents a valid nutritive support for the seedling and young plants, and provides a support to the pot frame increasing the mechanical performances.

The natural fibers used to reinforce the polymeric matrix were flexible and short fibers from tomato peels and seeds combined to more rigid, stiff and long fibers from hemp strands [120]. Hemp fibers, consisting of about 70% of cellulose, 15% of hemicelluloses, 5% of lignin and wax, and up to 10% of moisture [121] were on average 200 μm long, 10 μm wide and 5 μm thick. Tomato fibers were wastes of tomato-processing industry of Lycopersicon esculentum "S. Marzano", a typical Italian tomato cultivar [122]. The dry matter of tomato peels and seeds was mostly composed of about 50% of fibers and 20% of crude protein, while the remaining part was characterized by fats and carotenoids, such as lycopene and β-carotenoid. The cell walls of fibers contained cellulose, hemicelluloses and pectin, while starch was present as energetic and preserving plant source [123]. After the extraction of high added value bioactive molecules, such as polysaccharides, carotenoids and polyphenols, from peels and seeds, the residual dried fibers were used as received. Tomato fibers were on average 80 μm long, 10 μm wide, and 10 μm thick [122].

Polyglycerol was used as plasticizer; calcium chloride was used to crosslink sodium alginate.

The pots were made from three different compositions of biocomposites, soaking 50.0 g of aggregate of fibers in 100 ml of a 2% w/v sodium alginate water solution. The compositions of biocomposites were prepared varying the percentage of tomato and hemp fibers added to sodium alginate water solution: 100% of tomato fiber (coded ATH100), 90% of tomato fiber and 10% of hemp fiber (coded ATH90) and 70% of tomato fiber and 30% of hemp fiber (coded ATH70). The components were thoroughly mixed by means of a blender for 30 min at room temperature (cold process) and at a rate of 16 revolutions per minute (rpm). The paste was distributed inside the pot-shaped closed molds of a stainless steel device (Fig. 5.22). This device was fixed between the cold plates of a press where the molds were allowed to close, left for few minutes for completely fill the cavity with the wet paste and then opened, in this way providing the wet shaped pots. A following drying process, carried out in an oven at 40 °C under air flow for 24 h, allowed to remove the water content up to obtain a final constant weight. The pots were characterized by a height of 40 mm, an end base diameter of 40 mm, a top base diameter of 55 mm, a thickness of 4 mm, and a weight of 9.0 g (Fig. 5.23). Novel biodegradable pots for

Fig. 5.22 Stainless steel pots mold

Fig. 5.23 Experimental biodegradable pot

seedling transplanting were tested in real-field condition, inside a steel-constructed greenhouse from July 21, 2009 to October 30, 2009 at the experimental farm of the University of Bari in Valenzano (Bari), Italy, having latitude 41° 05′N, longitude 16° 53′E, altitude 85 masl.

The greenhouse was covered with an ethylene vinyl acetate (EVA) film, having a thickness of 200 μm and a total transmissivity coefficient in the solar radiation range (200–2500 nm) equal to 90.9%, a solar direct transmissivity coefficient equal to 56.7%, and a transmissivity coefficient in the long wave infrared range (LWIR, 7500–12,500 nm) equal to 22.5%. Greenhouse temperature was controlled by means of a natural and a forced ventilation system. Ventilation was provided automatically when inside air temperature exceeded 27 °C [102].

During the period from seeding to seedlings transplanting the biodegradable pots, together with commercial pots made of 100% polystyrene (coded PS) used as control, were allocated in a steel bench, filled with perlite for almost half of its

Fig. 5.24 Experimental biodegradable pots setting up

depth. The experimental design was a completely randomized design; the four treatments (ATH100, ATH90, ATH70, and PS) had four replicates with four pots per replicate. The pots were allocated inside the perlite for 2/3 of their height (Fig. 5.24). On July 21, 2009 three pepper seeds were put into the growing substrate in each pot. After emergence only one plant per pot was kept. The irrigation was provided by means of an intensive fogging system, realized with a PVC irrigation pipe placed all around the bench with plastic nebulizer nozzles, every 0.5 m, watering over 180°.

On August 25, 2009 the seedlings together with the biodegradable pots were transplanted in the soil, without transplant shock. The plants grown inside the PS containers were transplanted following the procedure of removing the plastic pots. Peppers were harvested on October 30, 2009.

All the biodegradable pots remained intact throughout the entire period of 35 days from seeding to seedlings transplanting, showing sufficient mechanical resistance to guarantee material functionality: the parts of the pots in contact both with the perlite and the air did not show visual biodegradation. In this period, the pots were subjected to a mean greenhouse air temperature of 31.6 °C and to a mean greenhouse relative humidity of 53.8% [102].

The biodegradable pots allowed the development of the root structure with good branching structure and the development of secondary branching, with which the plants up took water and nutrients. At the transplant, a significantly dense network of root hairs developed in ATH100 pots, a dense one in ATH70 pots and a less dense network in ATH90 pots while in PS pots, used as control, long roots dominated the root system, reducing overall root development (Fig. 5.25). The different pots significantly influenced the seedlings height. The highest seedlings were those grown inside the ATH100 pots (0.119 m) and inside PS pots (0.115 m); shorter seedlings were grown inside the ATH90 (0.099 m) and ATH70 pots (0.100 m) suggesting that the presence of the filler probably influenced the seedlings growth.

Fig. 5.25 Roots development in ATH100 and ATH90 biodegradable pots and in polystyrene pot (PS) at seedlings transplanting

After the transplant, the biodegradable containers degraded completely in 16 days allowing the passage of the roots through the containers walls. The roots spread in a radial fashion; no root rot or similar symptoms were observed.

At the end of crop cycle, the different pots influenced pepper height. The mean plant height was 0.76 m for the plants grown inside the ATH90 pots, 0.73 m inside the ATH100 and ATH70 pots, and 0.67 m for the control plants [102]. Besides the obvious benefits to the environment, biodegradable pots have better insulating qualities, excellent drainage qualities, and can have also a fertilizing effect; to these aim inorganic salts, i.e., nitrogen, phosphorus and potassium, will be added during the setting up of the biopots thus improving plant nutrition and health.

5.5   Conclusion

The production and management of crops need plastics for many applications. In particular, among agricultural practices, soil mulching, commonly employed to inhibit the weeds growth, to preserve the soil moisture and to increase the soil temperature, entails huge amounts of plastic films coming from petroleum sources. Moreover, in horticulture, the agricultural practice of transplanting, i.e., the removal of a plant from the pots where it has been growing to another location, soil or larger containers, involves the use of rigid containers made of fossil raw materials, such as polystyrene, polyethylene, and polypropylene.

Nevertheless, in the past two decades, the growing environmental awareness provoked by the high post-consumer plastic waste with the following problems related to their correct disposability, together with the economic issue concerning the un-renewability of source, since petroleum is the main feedstock for creating plastics, strongly encouraged researchers toward the use of biodegradable polymers. Hence, renewability and biodegradability have become key words in the debate over ecosustainable production and utilization of plastics. Recently, researchers have made strong efforts to identify new biopolymers coming from renewable sources as valid ecosustainable alternatives to petroleum-based plastic commodities.

In this chapter, the main research results and current applications concerning the biodegradable plastics, such as thermo-extruded Mater-Bi and sprayable water-born polysaccharides based coatings, have been described; in particular, a lineup of biopolymers coming from raw and renewable sources, such as polysaccharides, have been reported; particular attention has been given to the investigation of intrinsic chemicophysical properties of polysaccharides, responsible of the realization of dry water stable hydrogels, suitable for the formation of both soil protective geomembranes and transplanting biopots. Furthermore, a detailed description of the natural additives, fillers and cellulosic fibers included in the polymeric matrices, able to enhance the mechanical performance of coatings and cell trails have been provided, together with the outputs in the specific applications. Last but not least, the innovative approach of mulch forming by means of spray methodology is less labor demanding because it does not require manual operations such as the laying-out of plastic films on the trays or pots. Hence, the idea of using simple spray brushing machines instead of processing equipment will both reduce energy consumptions and decrease the development costs. Moreover, the use of natural products, some of them coming from renewable resources, and some of them from marine or agricultural wastes or by-products, will support the reducing of costs.

Finally, promoting biodegradable plastics, through widespread training and education will be more and more claimed if the benefits in terms of the environment and cost saving continue to be highlighted, emphasizing their both low environmental impact and industrial sustainability and commercial availability.

Acknowledgements The present work acknowledges the financial support from the Ministero dell'Università e della Ricerca Scientifica, Industrial Research Project "Development of green technologies for production of BIOchemicals and their use in preparation and industrial application of POLImeric materials from agricultural biomasses cultivated in a sustainable way in Campania region—BioPoliS" PON03PE_00107_1/1, funded in the frame of Operative National Programme Research and Competitiveness 2007–2013 D. D. Prot. n. 713/Ric. 29.10.2010.

The authors shared programming and editorial work equivalently within the areas of their expertise.

References

1. Romić D, Romić M, Borosić J, Poljak M (2003) Mulching decreases nitrate leaching in bell pepper (*Capsicum annuum* L.) cultivation. Agric Water Manag 60(2):87–97
2. Ramakrishna A, Tam HM, Wani SP, Long TD (2006) Effect of mulch on soil temperature, moisture, weed infestation and yield of groundnut in northern Vietnam. Field Crops Res 95:115–125
3. Singh R, Sharma RR, Goyal RK (2007) Interactive effects of planting time and mulching on 'Chandler' strawberry (*Fragaria* × *ananassa* Duch.). Sci Hortic 111:344–351
4. Tilander Y, Bonzi M (1997) Water and nutrient conservation through the use of agroforestry mulches, and sorghum yield response. Plant Soil 197:219–232
5. Green DS, Kruger EL, Stanosz LR (2003) Effects of polyethylene mulch in a short-rotation poplar plantation with weed-control strategies, site quality and clone. Forest Ecol Manag 173:251–260

6. Li FM, Guo AH, Wei H (1999) Effects of clear plastic film mulch on yield of spring wheat. Field Crop Res 63:79–86

7. Li FM, Wang P, Wang J, Xu JZ (2004) Effects of irrigation before sowing and plastic film mulching on yield and water uptake of spring wheat in semiarid Loess Plateau of China. Agr Water Manage 67:77–88

8. Malinconico M, Immirzi B, Santagata G, Schettini E, Vox G, Scarascia Mugnozza G (2008) An overview on innovative biodegradable materials for agricultural applications. In: Moeller HW (ed) Progress in polymer degradation and stability research (Chap. 3). Nova Science Publishers, Inc., pp 69–114

9. Johnson MS, Fennimore SA (2005) Weed and crop response to colored plastic mulches in strawberry production. Hort Sci 40:1371–1375

10. Stapleton JJ, Summers CG (2002) Reflective mulches for management of aphids and aphid-borne virus diseases in late-season cantaloupe (*Cucumismelo* L. var. cantalupensis). Crop Prot 21(10):891–898

11. Kasperbauer MJ, Loughrin JH, Wang SY (2001) Light reflected from red mulch to ripening strawberries affects aroma, sugar and organic acid concentrations. Photochem Photobiol 74 (1):103–107

12. Espì E, Salmeron A, Fontecha A, Garcia Y, Real AI (2006) Plastic films for agricultural application. J Plast Film Sheeting 22:85–102

13. Bilck AP, Grossmann MVE, Yamashita F (2010) Biodegradable mulch films for strawberry production. Polym Testing 29:471–476

14. Briassoulis D (2005) The effects of tensile stress and the agrochemical Vapamon the ageing of low density polyethylene (LDPE) agricultural films. Part I. Mechanical behavior. Polym Degrad Stab 88:489–503

15. Moreno MM, Moreno A (2008) Effect of different biodegradable and polyethylene mulches on soil properties and production in a tomato crop. Sci Hortic 116:256–263

16. Schettini E, Vox G, Stefani L (2014) Interaction between agrochemical contaminants and UV stabilizers for greenhouse EVA plastic films. Appl Eng Agric 30(2):229–239

17. Kasirajan S, Ngouajio M (2012) Polyethylene and biodegradable mulches for agricultural applications: a review. Agron Sustain Dev 32(2):501–529

18. Scarascia Mugnozza G, Sica C, Russo G (2011) Plastic materials in European agriculture: actual use and perspectives. J Agric Eng 3:15–28. http://www.esajournals.org/doi/full/10. 1890/EHS14-0018.1

19. Ren X (2003) Biodegradable plastic: a solution or a challenge? J Cleaner Prod 11:27–40

20. Picuno P (2014) Innovative material and improved technical design for a sustainable exploitation of agricultural plastic film. Polym Plast Technol Eng 53(10):1000–1011

21. Briassoulis D, Babou E, Hiskakis M, Scarascia G, Picuno P, Guarde D, Dejean C (2013) Review, mapping and analysis of the agricultural plastic waste generation and consolidation in Europe. Waste Manage Res 31(12):1262–1278

22. Picuno P, Sica C, Laviano R, Dimitrijević A, Scarascia Mugnozza G (2012) Experimental tests and technical characteristics of regenerated films from agricultural plastics. Polym Degrad Stab 97(9):1654–1661

23. Sica C, Dimitrijević A, Scarascia Mugnozza G, Picuno P (2015) Technical properties of regenerated plastic material bars produced from recycled agricultural plastic film. Polym Plast Technol Eng 54(12):1207–1214

24. http://www.marketsandmarkets.com/Market-Reports/agricultural-mulch-films-market741. html

25. Agricultural Films Market by Type (LLDPE, LDPE, HDPE, EVA, Reclaim), Applications (Greenhouse, Silage and Mulch) and Geography (North America, Asia-Pacific, Europe and RoW)—Global Trends & Forecast to 2020 Publishing Date: April 2015 Report Code: CH 3330

26. Lamont W (2004) Plasticulture: an overview. In: Lamont W (ed) Production of vegetables, strawberries, and cut flowers using plasticulture. Natural Resource, Agriculture, and Engineering Service (NRAES), Ithaca

27. Vox G, Schettini E, Stefani L, Modesti M, Ugel E (2008) Effects of agrochemicals on the mechanical properties of plastic films for greenhouse covering. Acta Hortic 801:155–162
28. Briassoulis D, Aristopoulou A, Bonora M, Verlodt I (2004) Degradation characterization of agricultural low density polyethylene films. Biosyst Eng 88(2):131–143
29. Briassoulis D (2005) The effects of tensile stress and the agrochemical Vapam on the ageing of low density polyethylene (LDPE) agricultural films. Part I. Mechanical behaviour. Polym Degrad Stab 88(3):489–503
30. Schettini E, Vox G (2012) Effects of agrochemicals on the radiometric properties of different anti-UV stabilized EVA plastic films. Acta Hortic 956:515–522
31. Schettini E, De Salvador FR, Scarascia Mugnozza G, Vox G (2011) Radiometric properties of photoselective and photoluminescent greenhouse plastic films and their effects on peach and cherry tree growth. J Hortic Sci Biotechnol 86(1):79–83
32. Kijchavengkul T (2010) Design of biodegradable aliphatic aromatic polyester films for agricultural applications using response surface methodology. Ph.D. dissertation, Michigan State University
33. Ingman M, Santelmann MV, Tilt B (2015) Agricultural water conservation in China: plastic mulch and traditional irrigation. Ecosyst Health Sustain 1(4):1–11
34. Jandas PJ, Mohanty S, Nayak SK (2013) Sustainability, compostability, and specific microbial activity on agricultural mulch films prepared from poly(lactic acid). Ind Eng Chem Res 52:17714–17724
35. Vox G, Schettini E, Scarascia Mugnozza G (2005) Radiometric properties of biodegradable films for horticultural protected cultivation. Acta Horticulturae 691(2):575–582
36. Malinconico M, Immirzi B, Massenti S, La Mantia FP, Mormile P, Petti L (2002) Blends of polyvinylalcohol and functionalised polycaprolactone. A study on the melt extrusion and post-cure of films suitable for protected cultivation. J Mater Sci 37:4973–4978
37. Kapanen A, Schettini E, Vox G, Itavaara M (2008) Performance and environmental impact of biodegradable films in agriculture: a field study on protected cultivation. J Polym Environ 16:109–122
38. Hayes DG, Dharmalingam S, Wadsworth LC, Leonas KK, Miles CA, Inglis DA (2012) Biodegradable agricultural mulches derived from biopolymers. In: Kishan AI, Khemani C, Scholz C (eds) Degradable polymers and materials, principles and practice. University of Alabama at Huntsville. ACS Books
39. Endres HJ, Siebert-Raths A (2011) Engineering biopolymers—markets, manufacturing, properties and applications. HanserPublishers, Munich
40. Scarascia Mugnozza G, Schettini E, Vox G, Malinconico M, Immirzi B, Pagliara S (2006) Mechanical properties decay and morphological behaviour of biodegradable films for agricultural mulching in real scale experiment. Polym Degrad Stab 91(11):2801–2808
41. Keshavarz T, Roy I (2010) Polyhydroxyalkanoates: bioplastics with a green agenda. Curr Opin Microbiol 13:321–326
42. Riggi E, Santagata G, Malinconico M (2011) Bio-based and biodegradable plastics for use in crop production. Recent Pat Food Nutr Agric 3:49–63
43. National Organic Standards (NOS) (2012) § 205.206 Crop pest, weed, and disease management practice standard. § 205.601(b)(2)(i–ii) Synthetic substances allowed for use in organic crop production
44. Briassoulis D, Dejean C (2010) Critical review of norms and standards for biodegradable agricultural plastics. Part I: Biodegradation in soil. J Polym Environ 18(3):384–400
45. ASTM International (2012) Standard specification for aerobically biodegradable plastics in soil environment. West Conshohocken, PA
46. ASTM D6400 (2004) International standard specification for compostable plastics. ASTM International, West Conshohocken
47. Tachibana Y, Maeda T, Ito O, Maeda Y, Kunioka M (2009) Utilization of a biodegradable mulch sheet produced from poly(lactic acid)/Ecoflex®/modified starch in mandarin orange groves. Int J Mol Sci 10:3599–3615

48. Sartore L, Vox G, Schettini E (2013) Preparation and performance of novel biodegradable polymeric materials based on hydrolyzed proteins for agricultural application. J Polym Environ 21(3):718–725

49. Schettini E, Vox G, De Lucia B (2007) Effects of the radiometric properties of innovative biodegradable mulching materials on snapdragon cultivation. Sci Hortic 112(4):456–461

50. Atkins EDT (ed) (1985) Polysaccharides: topics in structure and morphology. VCH Verlag, Weinheim. Iwata T (2015) Biodegradable and bio-based polymers: future prospects of eco-friendly plastics. Angew Chem Int Ed 54(11):3210–3215

51. Enas MA (2015) Hydrogel: preparation, characterization, and application: a review. J Adv Res 6(2):105–121

52. Schacht EH (2004) Polymer chemistry and hydrogel systems. J Phys Conf Ser 3:22–28

53. Ebara M, Kotsuchibashi Y, Narain R, Idota N, Kim YJ, Hoffman JM, Uto K, Aoyagi T (2014) Smart hydrogel. Smart Biomater 373:9–65

54. Avella M, Di Pace E, Immirzi B, Impallomeni G, Malinconico M, Santagata G (2007) Addition of glycerol plasticizer to seaweeds derived alginates: influence of microstructure on chemical–physical properties. Carbohyd Polym 69(3):503–51

55. Mormile P, Petti L, Rippa M, Immirzi B, Malinconico M, Santagata G (2007) Monitoring of the degradation dynamics of agricultural films by IR thermography. Polym Degrad Stabil 92:777–784

56. Santagata G, Malinconico M, Immirzi B, Schettini E, Scarascia Mugnozza G, Vox G (2014) An overview of biodegradable films and spray coatings as sustainable alternative to oil-based mulching films. Acta Hortic 1037:921–928

57. Šubarić D, Babić J, ĐurđicaAčkar VlastaPiližota, Kopjar M, Ljubas I, Ivanovska S (2011) Effect of galactomannan hydrocolloids on gelatinization and retrogradation of tapioca and corn starch Croat. J Food Sci Technol 3(1):26–31

58. Xiong JY, Narayanan J, Liu XY, Chong TK, Chen SB, Chung TS (2005) Topology evolution and gelation mechanism of agarose gel. J Phys Chem B 109:5638–5643

59. Qu X, Wirsén A, Albertsson AC (2000) Novel pH-sensitive chitosan hydrogels: swelling behavior and states of water. Polymer 41(12):4589–4598

60. Straccia MC, Romano I, Oliva A, Santagata G, Laurienzo P (2014) Crosslinker effects on functional properties ofalginate/N-succinylchitosan based hydrogels. Carbohydr Polym 108:321–330

61. Draget KI, Smidsrød O, Skjak-Braek G (2002) Alginates from algae. Biopolymers 6:215–244

62. Draget KI, Skiak-Braek G, Smidsrod O (1997) Alginate based new material. Int J Biol Macromol 21:47–55

63. Smidsrod O, Haug A, Larsen B (1966) The influence of pH on the rate of hydrolysis of acidic polysaccharides. Acta Chem Scand 20(4):1026–1034

64. Russo R, Abbate M, Malinconico M, Santagata G (2010) Effect of polyglycerol and the crosslinking on the physical properties of a blend alginate-hydroxyethylcellulose. Carbohydr Polym 82:1061–1067

65. Grasdalen H, Larsen B, Smidsrod O (1981) 13C-NMR studies of monomeric composition and sequence in alginate. Carbohydr Res 89:179–191

66. Draget KI, Skiak-Braek G, Stokke BT (2006) Similarities and differences between alginic acid gels and ionically crosslinked alginate gels. Food Hydrocolloid 20:170–175

67. Gomez-Diaz D, Navata JM (2004) Rheology of food stabilizers blends. J Food Eng 64:143–149

68. Grant GT, Morris ER, Rees DA, Smith PJC, Thom D (1973) Characterization of alginate composition and block-structure by circular dicroism. FEBS Lett 32:195–198

69. Morris ER, Rees DA, Thom D, Boyd J (1978) Chiroptical and stoichiometric evidence of a specific, primary dimerisation process in alginate gelation. Carbohydr Res 66:145–154

70. Chandrasekaran R, Puijaner LC, Joyce KL, Arnott S (1988) Cation interactions in gellan: an x-ray study of the potassium salt. Carboydr Res 181:23–40

71. Stokke BT, Draget KI, Smidsrod O, Yuguchi Y, Urakawa H, Kajiwara K (2000) Small-angle X-ray scattering and rheological characterization of alginate gels. 1. Ca-alginate gels. Macromolecules 33:1853–1863

72. Russo R, Malinconico M, Santagata G (2007) Effect of cross-linking with calcium ions on the physical properties of alginate films. Biomacromolecules 8:3193–3197

73. Chan LW, Lee HY, Heng PWS (2006) Mechanisms of external and internal gelation and their impact on the functions of alginate as a coat and delivery system. Carbohydr Polym 63 (2):176–187

74. Naidu BVK, Rao KSVK, Aminabhavi TM (2005) Pervaporation separation of water + 1,4-dioxane and water + tetrahydrofuran mixtures using sodium alginate and its blend membranes with hydroxyethylcellulose. A comparative study. J Membr Sci 260:131–141

75. Al-Kahtani AA, Sherigara BS (2014) Controlled release of diclofenac sodium through acrylamide grafted hydroxyethyl cellulose and sodium alginate. Carbohydr Polym 104 (15):151–157

76. Srivastava M, Kapoor VP (2005) Seed galactomannans: an overview. Chem Biodivers 2 (3):295–317

77. Khanna S, Tester RF (2006) Influence of purified konjac glucomannan on the gelatinisation and retrogradation properties of maize and potato starches. Food Hydrocolloids 20:567–576

78. Stanley NF (1995) Agars. In: Stephen AM (ed) Food polysaccharides and their applications. Marcel Dekker, New York, pp 187–204

79. Norziah MH, Foo SL, Karim AA (2006) Rheological studies on mixtures of agar (Gracilaria changii) and k-carrageenan. Food Hydrocolloids 20:204–217

80. Pillai CKS, Paul W, Sharma CP (2009) Chitin and chitosan polymers: chemistry, solubility and fiber formation. Prog Polym Sci 34(7):641–678

81. Muzzarelli RAA (1996) Chitin. In: Salamone JC (ed) The polymeric materials encyclopedia. CRC Press, Inc, Boca Raton

82. Peniche C, Argüelles-Monal W, Peniche H, Acosta N (2003) Chitosan: an attractive biocompatible polymer for microencapsulation. Macromol Biosci 3(5):11–20

83. Catanzano O, Straccia MC, Miro A, Ungaro F, Romano I, Mazzarella G, Santagata G, Quaglia F, Laurienzo P, Malinconico M (2014) Spray-by-spray in situ cross-linking alginate hydrogels delivering a tea tree oil microemulsion. Eur J Pharm Sci 66:20–28

84. Dutta J, Tripathi S, Dutta PK (2012) Progress in antimicrobial activities of chitin, chitosan and its oligosaccharides: a systematic study needs for food applications. Food Sci Technol Int 18(1):3–34

85. Romanazzi G, Feliziani E, Santini M, Landi L (2013) Effectiveness of postharvest treatment with chitosan and other resistance inducers in the control of storage decay of strawberry. Postharvest Biol Technol 75:24–27

86. Rinaudo M (2006) Chitin and chitosan: properties and applications. Prog Polym Sci 31 (7):603–632

87. Suyatma NE, Tighzert L, Copinet A (2005) Effects of hydrophilic plasticizers on mechanical, thermal, and surface properties of chitosan films. J Agric Food Chem 53:3950–3957

88. Siró I, Plackett D (2010) Microfibrillated cellulose and new nanocomposite materials: a review. Cellulose 17:459–494

89. Nordqvist D, Idermark J, Hedenqvist MS (2007) Enhancement of the wet properties of transparent chitosan-acetic-acid salt films using microfibrillated cellulose. Biomacromolecules 8:2398–2403

90. Sionkowska A, Planecka A, Kozlowska J, Skopinska-Wisniewska J, Los P (2011) Weathering of chitosan films in the presence of low- and high-molecular weight additives. Carbohydr Polym 84(3):900–906

91. Kalogeras IM, Brostow W (2009) Glass transition temperatures in binary polymer blends. J Polym Sci Part B Polym Phys 47:80–95

92. Nieto JM, Peniche-Covas C, Padron G (1991) Characterization of chitosan by pyrolysis-mass spectroscopy, thermal analysis and differential scanning calorimetry. Thermochimca Acta 176:63–68

93. Higuchi A, Iijima T (1985) DSC investigation of the states of water in poly(vinyl alcohol) membranes. Polymer 26:1207–1211

94. Cuperus FP, Bargeman D, Smolders CA (1992) Critical points in the analysis of membrane pore structures by thermoporometry. J Membr Sci 66:45–53

95. Evans MR, Hensley DL (2004) Plant growth in plastic, peat, and processed poultry feather fiber growing containers. HortScience 39(5):1012–1014

96. Evans MR, Karcher D (2004) Properties of plastic, peat, and processed poultry feather fiber growing containers. HortScience 39(5):1008–1011

97. Struve DK (1993) Effect of copper-treated containers on transplant survival and regrowth of four tree species. J Environ Hortic 11(4):196–199

98. Santagata G, Immirzi B (2009) Processability of biocomposite for agricultural application. In: Vasile C, Zaikov GE (eds) Environmentally degradable materials based on multicomponent polymeric systems (Chap. 15). CRC Press book, Boston, pp 575–585

99. Horinouchi H, Katsuyama N, Taguchi Y, Hyakumachi M (2008) Control of fusarium crown rot of tomato in a soil system by combination of a plant growth-promoting fungus, fusarium equiseti, and biodegradable pots. Crop Prot 27:859–864

100. Kowalska E, Wielgosz Z, Pelka J (2002) Use of post-life waste and production waste in thermoplastic polymer compositions. Polym Polym Compos 10(1):83–91

101. Yamauchi M, Masuda S, Kihara M (2006) Recycled pots using sweet potato distillation lees. Resour Conserv Recycl 47:183–194

102. Schettini E, Santagata G, Malinconico M, Immirzi B, Scarascia Mugnozza G, Vox G (2013) Recycled wastes of tomato and hemp fibres for biodegradable pots: physico-chemical characterization and field performance. Resour Conserv Recycl 70:9–19

103. Castronuovo D, Picuno P, Manera C, Scopa A, Sofo A, Candido V (2015) Biodegradable pots for poinsettia cultivation: agronomic and technical traits. Sci Hortic 197:150–156

104. Liew KC, Khor LK (2015) Effect of different ratios of bioplastic to newspaper pulp fibres on the weight loss of bioplastic pot. J King Saud Univ Eng Sci 27(2):137–144

105. Minuto G, Minuto A, Pisi L, Tinivella F, Guerrini S, Versari M, Pini S, Capurro M, Amprimo I (2008) Use of compostable pots for potted ornamental plants production. Acta Hortic 801:367–372

106. Trojanowsky J, Huttermann A (2002) Biodegradable materials and natural fibre composite in agriculture and horticulture. In: International symposium, 2–4 June 2002, Germany, pp 54–60

107. Sartore L, Bignotti F, Pandini S, D'Amore A, Di Landro L (2015) Green composites and blends from leather industry waste. Polym Compos (Article in Press). doi:10.1002/pc.23541

108. Šimkovic I (2008) What could be greener than composites made from polysaccharides? Carbohydr Polym 74:759–762

109. Mohanty AK, Misra M, Hinrichsen G (2000) Biofibres, biodegradable polymers and biocomposites: An overview. Macromol Mater Eng 276(277):1–24

110. Hatami-Marbini H, Pietruszczak S (2007) On inception of cracking in composite materials with brittle matrix. Comput Struct 85:1177–1184

111. Immirzi B, Santagata G, Vox G, Schettini E (2009) Preparation, characterisation and field testing of a biodegradable sodium alginate-based spray mulch. Biosyst Eng 102:461–472

112. Coviello T, Matricardi P, Marianecci C, Alhaique F (2007) Polysaccharide hydrogels for modified release formulations. J Control Rel 119:5–24

113. Ahmad NNR, Fernando WJN, Uzir MH (2015) Parametric evaluation using mechanistic model for release rate of phosphate ions from chitosan-coated phosphorus fertilizer pellets. Biosyst Eng 129:78–86

114. Lewin LM, Pearce EM (1985) Handbook of fibre science and technology. Fibre Chemistry, vol IV. Marcel Dekker, New York

115. Wambua P, Ivens U, Verpoest I (2003) Natural fibres: can they replace glass in fibre-reinforced plastics? Compos Sci Technol 63:1259–1264

116. Cyras VP, Martucci J, Iannace S, Vázquez A (2001) Influence of the fibre content and the processing conditions on the flexural creep behaviour of sisal–PCL–starch composites. J Thermoplast Compos Mater 14:1–13
117. Hepworth DG, Bruce DM (2003) The mechanical properties of a composite manufactured from non-fibrous vegetable tissue and PVA. Compos A 31:283–285
118. Mohanty AK, Mubarak AK, Hinrichsen G (2000) Surface modification of jute and its influence on performance of biodegradable jute-fabric/biopol composites. Comp Sci and Tech 60:1115–1124
119. Shinde UA, Nagarsenker MS (2009) Characterization of gelatin-sodium alginate complex coacervation system. Ind J Pharm Sci 71:313–317
120. Ashori A, Nourbakhsh A (2010) Bio-based composites from waste agricultural residues. Waste Manage 30:680–684
121. Amar KM, Manjusri M, Lawrence TD (2005) Natural fibres, biopolymers, and bio-composites. CRC Press, Tailor & Francis
122. Tommonaro G, Poli A, De Rosa S, Nicolaus B (2008) Tomato derived polysaccharides for biotechnological applications: chemical and biological approaches. Molecules 13:1384–1398
123. Knoblich M, Anderson B, Latshaw D (2005) Analyses of tomato peel and seed byproducts and their use as a source of carotenoids. J Sci Food Agric 85:1166–1170

Chapter 6
Standards for Soil Biodegradable Plastics

Demetres Briassoulis and Francesco Degli Innocenti

Abstract The main standard test methods for biodegradation of plastics in soil (ISO 17556, ASTM D5988, NF U52-001 and UNI 11462) determine the rate of biodegradation under normalized conditions. The standard testing procedures are designed to determine the inherent biodegradability of plastics in soil under optimal controlled conditions that may not be necessarily representative of any specific environmental conditions but they ensure repeatability. The normalized conditions defined by the standard test methods differ in several aspects. A comparative analysis is presented. Besides the biodegradation test methods, pass levels and a time frame also need to be defined in order to determine whether bio-based products will biodegrade sufficiently under soil conditions. There is currently no European or international specification that defines criteria for biodegradation of bio-based products in soil. Criteria for biodegradation of materials used in agriculture and horticulture are only defined in standard specifications NF U52-001 and UNI 11462, together with criteria for environmental safety. However, the evaluation of the biodegradation in soil is not obligatory in the French specification. The main requirements for mulching films are: (1) biodegradation at least 90% within 24 months; (2) material shall not contain heavy metal, no ecotoxicological effects. The same requirements have been adopted by the USDA-AMS National Organic Program (NOP) for mulching films allowed for organic crop production. The constraints, gaps, and limitations of existing relevant testing methods and the new developments are identified and analyzed in this chapter. Functional barriers with respect to standards and labeling for soil biodegradable plastics are analyzed.

D. Briassoulis (✉)
Agricultural Engineering Department, Agricultural University of Athens,
75 Iera Odos, 11855 Athens, Greece
e-mail: briassou@aua.gr

F. Degli Innocenti
Novamont S.p.A., via Fauser 8, 28100 Novara, Italy
e-mail: fdi@novamont.com

© Springer-Verlag GmbH Germany 2017
M. Malinconico (ed.), *Soil Degradable Bioplastics for a Sustainable Modern Agriculture*, Green Chemistry and Sustainable Technology,
DOI 10.1007/978-3-662-54130-2_6

Keywords Standard testing methods · Standard specification · Soil biodegradable plastics · Mulching films · Biodegradation · Bio-based films · Oxo-degradable films · Ecotoxicological effects

6.1 Introduction

Almost half of the plastics worldwide are used for disposable applications, including packaging, mulching films, and other disposable consumer items [1]. Agricultural applications in Europe in 2012 have a share of 4.2% of plastics demand [2]. Waste generated from agricultural plastics is estimated at 5–6% of the total plastic waste in Europe. An estimated 2–3 million tons of plastics, mainly polyethylene, are used each year in agricultural applications worldwide [3]. Half of this amount is used in protected cultivations (greenhouses, mulching, small tunnels, etc.). However, the expanding use of plastics for protected horticulture has a major negative consequence: handling of plastic waste and the associated environmental impact. Only a small percentage of agricultural plastic waste is currently recycled (though this percentage varies widely from region to region or country and depends on the plastic category, e.g., most silage films and bale wraps are recycled) [3]. The handling of agricultural plastic waste, unlike municipal plastic waste, requires special management schemes and infrastructure organized at the level of rural areas [4]. A large portion of this waste is left in the fields, buried or burnt uncontrollably by the farmers, releasing harmful substances with the associated negative consequences on the environment and possibly for the safety of the food produced (including aesthetic pollution and landscape degradation, degradation of soil quality characteristics; the release of harmful substances with negative consequences on the environment and human health and possible danger to the safety of the food produced) [3].

143,000 t of plastic mulch was disposed of in the US in 2004, either in landfills or burned on site [5]. Similar quantities of plastic mulching film waste were generated in Europe [6]. Burning or burying polyethylene mulch in fields is associated with undesirable environmental and public health impacts (e.g., release of airborne toxic substances, including dioxins, and irreversible soil contamination) and is illegal in both EU and the US [3, 7]. The main reasons for the mismanagement of plastic mulching (and other agricultural) film waste are the lack of cost-efficient systematic disposal techniques and management systems available to the growers along with the high labor cost for the proper collection of these plastic wastes following the end of the cultivation season [3, 8]. Costs for polyethylene mulch disposal (not including labor) were estimated at up to 250 $/ha [5]. In addition, recycling is not considered as a cost-efficient and/or technically feasible solution, for dirty thin mulching or low-tunnel films. The alternative for these materials is the costly option of energy recovery.

All these reasons have led to a great interest in biodegradable plastics. Plastics designed to biodegrade in contact with soil at the end of their commercial lives can be used in several applications. For example, biodegradable plastics can be used in products that are intentionally used in soil contact, e.g. agricultural mulching films, and in products where soil is the inevitable final location e.g. golf tees, plastic "clays" for clay pigeon shooting, etc.

On the other hand, soil quality is a major concern. Soil as an environment for biodegradation is considered very important since it is the center of great biological activity and the precious resource for food and fiber production, but also the medium in which soil-biodegradable plastics are deliberately disposed of. For bio-based materials which degrade in soil, the rate of degradation can vary considerably, depending not only on the molecular structure of the material, but also on soil characteristics and soil conditions such as temperature, water, and oxygen availability which vary widely and influence microbial activity. Materials that are not soil-biodegradable may result in contamination of the soil.

The interest in the development of biodegradable plastics to be safely used in agricultural applications has consequently led to the development of intense work at standardization level. Standardization plays a crucial role in biodegradable plastics. Biodegradability is not a characteristic that can be noted directly by consumers who must rely on the producers' claims. It is therefore obvious that in order to guarantee market transparency, tools are needed to provide a clear connection between declarations used as advertising messages and the actual biodegradability of the products. The important work on standardization carried out over the past 15 years in the bioplastics and biodegradable and compostable packaging sector is an example of the role played by standardization in society and to support innovation [9–11].

Standardization is, according to the ISO (International Organization for Standardization) [12], the "*activity of establishing, with regard to actual or potential problems, provisions for common and repeated use, aimed at the achievement of the optimum degree of order in a given context*" [13].

A fundamental task of standardization is to provide unequivocal terminology to describe the sector of interest. Before starting the close examination of the available relevant standards, some clarification on terminology is necessary for those who are not experts in this field. The terms "standard" and "norm" (or better, "technical norm") are synonymous. For the sake of clarity the term "standard" is used throughout the chapter to indicate a technical norm and "regulation" to indicate legislative acts.

Two terms are frequently used as synonyms (also by experts): biodegradability and biodegradation. However, biodegradability refers to a potentiality (i.e., the ability to be degraded by biological agents) while biodegradation refers to a process, which occurs under certain conditions, in a given time, with measurable results. The inherent biodegradability of a plastic is inferred by studying an actual biodegradation process under specific laboratory conditions, and the conclusion that the plastic is biodegradable (i.e., it can be biodegraded) in a specific environment can be drawn from the test results [14].

The biodegradation level reached by a plastic when exposed to a soil inoculum can be measured by means of a laboratory standard test method. On the other hand, the biodegradability of a plastic is defined by the so-called "standard specifications."

It must be pointed out that standard test methods do not set a pass level, since they are merely tools to measure a characteristic (i.e., biodegradation) under controlled conditions. Conversely, the "standard specifications" set pass levels and criteria that can be used in order to designate a plastic as "biodegradable" in soil on the basis of the test results obtained by applying specific test methods.

Biodegradation of a plastic material is studied under specific laboratory conditions because it is only under these conditions that reliable and reproducible scientific data can be gathered. The same approach has also been followed for chemicals (e.g., detergents).

Needless to say, the biodegradation behavior of biodegradable plastics under real soil conditions can deviate from the laboratory results and this can cause questions on the transferability of results in different soil types. The standard testing procedures are designed to determine the inherent biodegradation characteristics under an optimal controlled biodegradation process that may not be representative of the biodegradation of the specific bio-based materials/products under specific soil conditions but they ensure repeatability.

Besides a biodegradation test method, pass levels and a time frame also need to be defined in order to determine that a plastic product will biodegrade sufficiently under soil conditions. There are currently no European or international specifications to define criteria for the biodegradation of bio-based products in soil, though national specifications have been developed in France and in Italy.

The constraints, gaps, and limitations of existing relevant testing methods and the new standardization developments are identified and analyzed in this chapter. Needless to say, the reader who wants to apply a specific standard should obtain a copy of the original standard from the relevant Standardization Body.

6.2   Current Standards for Soil Biodegradable Plastics

6.2.1   Standard Methods for Testing Biodegradation of Plastics in Soil

The main laboratory standard test methods for testing biodegradation of plastics in soil are ISO 17556-12 [15] and ASTM D5988-12 [16]. CEN has adopted the ISO standard test method in the form of EN ISO 17556 standard [17].

These test methods determine biodegradation of plastics in soil under normalized laboratory conditions. Additional standards determine biodegradation of organic chemicals in soil (e.g., ISO 11266 [18]). Differences between the normalized conditions of these standards were identified and analyzed within the framework of the KBBPPS project [19, 20]. These differences concern several parameters of the

Table 6.1 Standard methods for testing biodegradation of plastic materials and products in soil

Current versions of standards	Title
American Society for Testing and Materials International (ASTM)	
ASTM D 5988-12	Standard test method for determining aerobic biodegradation of plastic materials in soil
International Organization for Standardization (ISO)	
ISO 17556-2012*	Plastics—determination of the ultimate aerobic biodegradability in soil by measuring the oxygen demand in a respirometer or the amount of carbon dioxide evolved
National Normalization Organizations (AFNOR, UNI)	
NF U52-001 February 2005	Biodegradable materials for use in agriculture and horticulture mulching products—requirements and test methods (test methods described in Annex F—Évaluation de la biodégradabilité aerobe dans le sol des produits par dégagement du dioxide de carbone)
UNI 11462:2012	Plastic materials biodegradable in soil—types, requirements and test methods (test methods described in Annex A—Determinazione della biodegradabilità aerobica in suolo mediante misurazione della quantità di CO_2 sviluppata usando un analizzatore ad infrarossi)

(*) Also EN ISO 17556

testing methods: (a) soil medium (natural soils from specified locations, laboratory mixture of soils, mixture of natural soil and mature compost or "standard soil"); (b) test sample (intact, cut into pieces or pulverized); (c) soil pH (natural or adjusted); (d) C/N ratio (natural or adjusted to 10:1–20:1 to the added organic C in the test sample or at least to 40:1 for sample organic C to soil N). In addition, there are several factors influencing biodegradation in soil which may be the reason for the poor reproducibility observed with these tests in some cases [15]. Soil water content (WC) measured as a percentage of the Water Holding Capacity (WHC) of the soil is one such important parameter affecting biodegradation.

The most important currently available standard testing methods for assessing biodegradation of plastics in soil are presented in Table 6.1. The basic principles and provisions of these testing methods are analyzed in this section with the aim of identifying and clarifying the constraints, gaps, and limitations and providing an update on the proposed modifications.

In addition, standard testing methods and guidelines are available for assessing biodegradation of chemicals in soil (Table 6.2). Relevant information for biodegradation of plastics in soil was presented earlier in [21] while a more detailed and updated comparison, including plastics and chemicals, was presented recently within the framework of the KBBPPS project [19, 22].

6.2.1.1   ASTM Standards for Testing Biodegradation of Plastics in Soil

The ASTM standard testing method (ASTM D 5988-12) [16], is *"applicable to all plastic materials that are not inhibitory to the bacteria and fungi present in soil."*

Table 6.2 Standard methods for testing biodegradability of chemicals in soil

Current versions of standards	Title
International Organization for Standardization (ISO)	
ISO 11266-1994	Soil quality guidance on laboratory testing for biodegradation of organic chemicals in soil under aerobic conditions
OECD guidelines	
304A	Inherent biodegradability in soil
307	Aerobic and anaerobic transformation in soil

This test method evaluates the degree of aerobic biodegradation of plastics "*by measuring evolved carbon dioxide as a function of time that the plastic is exposed to soil.*" The rate of biodegradation of plastic materials, including formulation additives, is measured against that of a positive reference material in an aerobic environment, thus allowing for an estimation of the degree of biodegradability and the biodegradation time under aerobic soil conditions.

This standard makes a significant clarification concerning the biodegradability specification: the results of the ASTM D 5988-12 test method (but applicable for any test method) are reported as the percentage of net CO_2 evolved for both the test and reference samples at the completion of the test and "*may not be used for unqualified "biodegradable" claims*" and "*may not be extrapolated beyond the actual duration of the test*" [16].

Soil medium

- The soil should be natural and fertile, preferably "sandy loam" collected from fields and forests not exposed to pollutants.
- The laboratory mixture is made of equal soil samples from three or more locations.
- The soil is sieved so that soil particle size is less than 2 mm.
- The soil used is fresh or reactivated in the case of air-dried or frozen soils.
- It is also acceptable to use a mixture of natural soil and mature compost as a test matrix.

Technical characteristics

- Measurement of evolved CO_2 as a function of time of exposure.
- Determination of the amount of test material based on the ratio of 200–1000 mg carbon of the test material for 500 g soil.
- Technical specifications: room temperature (e.g. 20–28 ± 2 °C); soil medium conditions: pH 6–8; moisture content to 80–100% of the moisture holding capacity (MHC) of the soil C:N ratio (assuming it refers to the test sample) is adjusted to a value between 10:1 and 20:1 by weight (e.g., with ammonium phosphate solution).

Validation criteria

- A control biodegradable substance (e.g. cellulose) must also be tested, in order to check the activity of the soil. If the biodegradation rate of the control sample at 6 months is less than 70%, the test is considered invalid and should be repeated with fresh soil.
- The measured CO_2 or the BOD values from the blanks at the end of the test or at the plateau phase are within 20% of the mean. Otherwise the test is considered invalid and should be repeated with fresh soil.

Applicability

- All plastic materials that are not inhibitory to the bacteria and fungi present in soil.

Equivalence

- This test method is equivalent to ISO 17556 [15].

6.2.1.2 ISO Standards for Testing Biodegradation of Plastics and Chemicals in Soil

ISO 17556:2012: The standard test method ISO 17556:2012 [15] "*specifies a method for determining the ultimate aerobic biodegradability of plastic materials in soil by measuring the oxygen demand in a closed respirometer or the amount of CO_2 evolved.*" Materials that may be tested include natural and/or synthetic polymers, copolymers or their mixtures, plastic materials containing additives and water-soluble polymers. The method is not applicable to materials inhibiting the activity of the microorganisms present in the soil. The test method "*is designed to yield an optimum degree of biodegradation by adjusting the humidity of the test soil*" [15].

Soil medium

- Natural soil from fields and/or forests may be used as an inoculum to simulate biodegradation in a specific natural environment.
- The soil is sieved so that soil particle size is less than 5 mm, preferably 2 mm.
- A standard soil may also be used to investigate the potential biodegradation of a test material in clayey or loamy soils.
- The standard soil is composed of industrial quartz sand, clay, natural soil and mature compost. Also, specific salts are added to the soil preferably when adjusting the water content.

Technical characteristics

- Measurement of BOD in a closed respirometer or the amount of CO_2 evolved.
- The test period should typically not exceed 6 months. If significant biodegradation is observed beyond 6 months, the test may be extended up to 24 months.

- Determination of the amount of test material based on the ratio of 100–300 mg of test material to 100–300 g of soil (recommended 200 mg test material for 200 g soil). In the case of measuring the evolution of CO_2 it is recommended to use a higher amount of test material, e.g. 2500 mg of test material for 200 g soil, to offset the fluctuations of the CO_2 produced by the blank control soil sample.
- Technical specifications: Room temperature: constant to within ±2 °C in the range between 20 and 28 °C, preferably 25 °C; Ratio C/N: 40/1 for organic C of test or reference material to nitrogen in the soil; Optimum water content of soil between 40 and 60% of total water holding capacity; pH: 6–8.

Validation criteria

- The degree of biodegradation of the reference material [microcrystalline cellulose powder, ashless cellulose filters or poly(-hydroxybutyrate)] is more than 60% at the plateau phase or at the end of the test.
- The measured CO_2 or the BOD values from the blanks at the end of the test or at the plateau phase are within 20% of the mean.
- Otherwise the test must be repeated using fresh preconditioned or pre-exposed soil.

Equivalence

- This test method is equivalent to ASTM D 5988.

Interpretation of the results

- Information on the toxicity of the test material may be useful in the interpretation of test results showing a low biodegradability.

ISO 11266-1994: The ISO 11266-1994 standard test method [17] does not describe a specific test method but offers guidance "*on the selection and conduct of appropriate test methods for the determination of biodegradation of organic chemicals in aerobic soils.*" Even though this standard does not concern biodegradation of plastics in soil, it is still briefly presented as it relates to aerobic biodegradation of materials in soil. Usually, a radiolabeled compound is used during laboratory testing, allowing the determination of the rate of disappearance of the test compound.

Soil medium

- If possible soil samples should be collected from the site under consideration.
- Alternatively the soil collected should have comparable properties.
- The field history of the test soil should be recorded including any agricultural machinery activities, application of agrochemicals, etc.

Technical characteristics

- Test materials are pure compounds (chemical purity >98%). The presence of any additives (e.g., formulation ingredients) has to be considered.

- The test chemical may be added in water (depending on their solubility in water), in organic solvents or directly as a solid (e.g., mixed in quartz sand).
- Determination of oxygen demand in a closed respirometer (based on ISO 9408 [23]). Use of radiolabeled compound or analytical approaches to determine the rate of test compound disappearance and "*the formation of metabolites, CO_2, other volatiles and non-extractable residue.*" The disappearance of the test compound and the identification of metabolites can also be followed using analytical methods.
- The test duration is not recommended to be longer than 120 days.
- Test substance: concentration depends on the experimental objectives.
- Room temperature: constant to within ±2 °C in the range between 25 and 35 °C, while for temperate zones the acceptable temperature range is between 10 and 25 °C.
- Ratio C/N: Properties found in natural soil.
- Optimum water content of soil between 40 and 60% of total water holding capacity.
- pH: pH found in natural soil.

6.2.1.3 Laboratory Test Method of Annex F of NF U52-001

The French standard specification NF U52-001 (2005) [24] determines the biodegradability of agricultural films in soil [20] and includes in Annex F a laboratory test method for soil biodegradation.

Soil medium

- A natural soil is to be used and the origin recorded in the test report (note: information about biodegradation in fresh water and compost is given in Annexes E and G).

Technical characteristics

- C:N should be adjusted to 10:1 up to 20:1 of organic carbon in the sample to total N in the soil (by addition of monohydrate of ammonium phosphate to the soil).
- pH: 6–8; water content at 80% of saturation; natural soil, sieved <2 mm with organic C < 2%.
- Sample containing between 200 mg and 1 g of organic carbon in 500 g of soil substrate; sample is added as fragments (with a length of 1–2 cm) or as powder.

Validation criteria for soil biodegradability testing

- Degree of biodegradation of microcrystalline cellulose (reference material) in the soil is more than 70% at plateau phase or at the end of a 6-month period.
- Replicate between the tests of the same material should not present more than 20% relative variation.

Applicability

- Biodegradable mulching films for agriculture and horticulture.

6.2.1.4 Laboratory Test Method in Annex A UNI

The Italian standard **UNI 11462:2012** [25] defines the requirements of biodegradability and ecotoxicity to be met by the polymers and plastics that are used to prepare manufactured articles mostly applied in the agriculture sector and which at the end of their use are left on or in the soil and in this environment are completely biodegraded without leaving toxic residues. A test method for soil biodegradation is described in Annex A.

Soil medium

A natural soil taken from a meadow, cultivated field or forest should be used, discarding the superficial layer. pH and water retention capacity are measured and adjusted to predefined ranges. To avoid a limitation of essential nutrients, which are necessary for a fast biodegradation process, the soil sample is supplemented with a mixture of salts and a small amount of compost. Alternatively, a "synthetic" soil can be prepared on the basis of a standard recipe.

Technical characteristics

- The system is based on a continuous airflow and on the measurement of the evolved carbon dioxide with an Infrared monitor.
- Technical specifications: room temperature (e.g. 21–28 °C); suggested airflow 5–10 L/h.

Validation criteria

The test is considered valid if the degree of biodegradation of the reference material is higher than 60% after 3 months and the biodegradation values of the replicas do not differ by more than 10%.

Equivalence

- This test method is considered as an alternative to ISO 17556 [15].

6.2.2 Standard Specifications and Regulation on Biodegradability of Plastics in Soil

There are only a few standard specifications and certification schemes defining biodegradability of plastic materials and products in soil. These are presented in Table 6.3.

Table 6.3 Standard specifications for determining biodegradability of materials and products in soil

French Normalization Organization (AFNOR)	
NF U52-001 February 2005	Biodegradable materials for use in agriculture and horticulture—mulching products—requirements and test methods
Italian Normalization Organization (UNI)	
UNI 11462:2012	Plastic materials biodegradable in soil—types, requirements and test methods
UNI 11495:2013	Biodegradable thermoplastic materials for use in agriculture and horticulture Mulching films—requirements and test methods
Other Specifications	
–	

6.2.2.1   French Standard NF U52-001 for Biodegradable Agricultural Films in Soil

The French Standard NF U52-001 (2005) [24] defines biodegradability of plastics in soil based on the results of the corresponding testing method.

According to the NF U52-001 standard, tested products are classified according to their expected lifetime.

Biodegradability testing requirements and criteria

- Tests for biodegradation of mulching films in soil done in three media as summarized in Table 6.4, which also presents the specifications set for each medium: water, soil and compost.
- The time and the minimum biodegradation percentage must be obtained.
- Minimum biodegradation percentage (%): 90 (water), 60 (soil), 90 (compost).
- Time (months): 6 (water), 12 (soil), 6 (compost).
- The minimum biodegradation percentage should be reached for at least two of the three media for validation of the biodegradability of the mulching film. However, the soil medium does not necessarily have to be one of the two media to be tested for the validation of its biodegradability in soil.

Environmental safety testing requirements and criteria

- Threshold limits are set for heavy metal, fluorine, PCB (polychlorinated biphenyl), and PAH (polycyclic aromatic hydrocarbon) content.
- Ecotoxicity tests:

Table 6.4 Biodegradability specifications set by the French Standard NF U52-001 (2005) [24]

Parameters	Fresh water	Soil	Compost
Biodegradability (%) in comparison to cellulose reference	90	60	90
Time (months)	6	12	6

- Emergence and growth of 1 mono and 1 dicotyledonous plant (ISO 11269-2 [26])
- Acute earthworm toxicity test (FD X 31-251 [27])
- Growth inhibition test with *Pseudokirchneriella subcapitata* (NF T 90-375 [28]).

Requirements with regard to labeling:
The Packaging label should indicate:

- Product conforms to NF U 52-001
- Name and address of producer of the product
- Commercial name or reference of the product
- Composition: Families of the three principle components
- Length
- Width
- Thickness
- Apparent density
- Class of material: A, B, C, D, E
- Final disposal: burying/composting
- Fabrication date and lot No
- Storage condition in original packaging (temperature, humidity, light, etc.).

NOTE: The expiration date of the product under optimal storage conditions and in the original packaging (month, year) may be mentioned.

On the Mulching Film:

The commercial name or reference number of the material should be printed on the seam if possible, otherwise it should be printed on the tube around which it is rolled.

NOTE: The expiration date of the product under optimal storage conditions (month, year) may be mentioned.

6.2.2.2   Italian Standard Specification UNI 11462 for Biodegradable Plastic Materials in Soil

UNI (the Italian Standardization Body) published the standard UNI 11462:2012 "Plastic materials biodegradable in soil—Types, requirements and test methods" [25] in 2012. A year later, the standard UNI 11495 (Biodegradable thermoplastic materials for use in agriculture and horticulture Mulching films—Requirements and test methods) was published with specific requirements regarding the characteristics of mulch films [29]. Since the biodegradability and ecotoxicity requirements of UNI 11462 are identical to those of UNI 11495, focus is placed on the former.

The Italian standard defines the biodegradability (see Table 6.5) and ecotoxicity requirements that plastic materials used to make products whose end-of-life is biodegradation in soil must possess.

Table 6.5 Biodegradability specifications set by Italian Standard UNI 11462 (2012) [25]

Parameters	Soil
Biodegradability (%) in comparison to cellulose reference	90
Time (months)	24

Biodegradability testing requirements and criteria

- Biodegradation is assessed with ISO 17556 or the test described in Annex A.
- Minimum biodegradation level must be 90% absolute or relative to the reference material.
- Time: 2 years maximum.

Environmental safety testing requirements and criteria

- Specific threshold limits are set for metals (As, Cd, Cr, Cu, F, Hg, Ni, Mo, Pb, Se, Zn, Co).
- Ecotoxicity tests:

  - Germination index, following UNI 10780 Annex K [30]
  - Growth index, following UNI 10780 Annex L
  - Acute toxicity test on earthworms, following ISO 6341 [31].

Requirements with regard to labeling

- Reference to the UNI standard
- Commercial name or reference of the product
- Trademark and name of producer
- Fabrication date
- Designation.

6.2.2.3 CEN and ISO Standards Specifications for Aerobically Biodegradable Plastics in Soil

No standard specifications have been developed yet at a European or international level.

6.2.2.4 Belgian Royal Decree for Acceptance of Compostable and Biodegradable Plastic Materials

The Belgian royal decree [32] defines three end-of-life management options for products: compostable, home compostable and biodegradable. It determines the requirements and standards that have to be fulfilled by each of these product categories with regard to its biodegradability and environmental safety [19].

Biodegradability testing requirements and criteria: The materials should conform to French specification NF U 52-001 for biodegradability of agricultural films in soil. A minimum rate of biodegradation of 90% absolute or relative (reference material: microcrystalline cellulose) should be achieved within 24 months [32].

Environmental safety testing requirements and criteria: The ecotoxicity tests required include [32]: OECD 208 test (refer to next section) in combination with the ecotoxicity tests described in standard EN 13432.

6.2.3 Standards for Measurement of Ecotoxicity of Chemicals in Soil Media

An overview of guidelines developed by the OECD, international standards and an ASTM standard to evaluate ecotoxicity of chemicals in soil media is given in Table 6.6.

The OECD methods [33–36] are designed to assess various levels of toxicity effects in short and long-term tests in terrestrial systems. Likewise the ecotoxicity essays of ISO for terrestrial systems [26, 37, 38] and ASTM [39]. The aim of measuring the ecotoxicity of chemicals in terrestrial systems is to evaluate the

Table 6.6 Overview of OECD guidelines and standard tests with regard to ecotoxicity of chemicals in soil

OECD guidelines	
Guidelines/Standards	Description
OECD 207 (4-4-1984)	Earthworm, acute toxicity tests
OECD 208 (19-7-2006)	Terrestrial plant test: seedling emergence and seedling growth test
OECD 222 (13-4-2004)	Earthworm reproduction test (*Eisenia fetida/Eisenia Andrei*)
OECD 317 (22-7-2010)	Bioaccumulation in terrestrial *Oligochaetes*
International Organization for Standardization (ISO)	
ISO 11268-1:2012	Soil quality—effects of pollutants on earthworms (*Eisenia fetida*)—part 1: determination of acute toxicity using artificial soil substrate
ISO 11269-2:2012	Soil quality—determination of the effects of pollutants on soil flora—part 2: effects of chemicals on the emergence and growth of higher plants
ISO 22030:2005	Soil quality—biological methods—chronic toxicity in higher plants
American Society for Testing and Materials International (ASTM)	
ASTM E1676—12	Standard guide for conducting laboratory soil toxicity or bioaccumulation tests with the Lumbricid Earthworm *Eisenia Fetida* and the Enchytraeid Potworm *Enchytraeus albidus*

adverse effects of contaminants associated with soil (chemicals, biomolecules, compounds, or additives released during degradation of materials in soil) on earthworms (Family *Lumbricidae*) and pot worms (Family *Enchytraeidae*) or on higher plants seedlings.

6.2.4   Labeling

6.2.4.1   Public Awareness and Education

Not all bio-based plastics are biodegradable and those which are biodegradable may effectively undergo biodegradation only under specific conditions but not in other conditions. According to the report of DG Environment on Plastic Waste in the Environment (2010) [40], *"clear certification and labeling schemes are needed to ensure the public understand the terms biodegradable, compostable and eco-friendly."* In the same report it is proposed that any targets on bio-based plastics (e.g., the development and application areas of bio-based plastics, performance, etc.) *"should be combined with a labeling system and initiatives to increase public awareness and education. Labeling of plastic parts with the type of polymer they contain could also help in sorting for recycling and reuse"* [40]. Certification and labeling is carried out by independent certification institutes and laboratories.

6.2.4.2   OK Biodegradable SOIL

A biodegradable material should be labeled by recognized certification organisms as being biodegradable in a specific medium condition [3] in accordance with the relevant standard testing methods and specifications described earlier. Thus in the case of mulching films, they may be labeled in accordance with the French NFU52-001 [24] specifications referring to the product "biodegradable mulching film" and are required to pass the test specifications in at least two media (one of which must be soil; proposed amendment for this standard in [20]).

The Belgian certification institute *VINÇOTTE* [41] has established a certificate for use of the "OK biodegradable SOIL" conformity mark (label) for products used in horticultural and agricultural applications. The label is applicable to the finished products as well as to all raw materials, components, and intermediate products. A number of standard testing methods are used and specific requirements are applied in evaluating the biodegradation and environmental safety of the products tested. These include 90% biodegradation (absolute or relative to a reference material) of the complete product/material (or for each organic constituent present in more than 1% of the material) within a maximum period of 2 years measured by the ISO 17556, ISO 11266, or ASTM D5988 test methods. The total proportion of organic constituents, not tested on biodegradation, may not exceed 5%. The certification includes assessment of ecotoxicity. No assessment of ecotoxicity is necessary for constituents accounting for less than 0.1% of the dry weight of the material or product, provided that the total percentage of these constituents does not exceed 0.5% of the dry weight of this material or product.

6.2.4.3 Bio-Based Mulch Films for Organic Farming

The U.S. Department of Agriculture's National Organic Program (NOP) has issued a final rule amending the National List of Allowed and Prohibited Substances (National List), which governs the use of synthetic and nonsynthetic substances in organic crop production and processing [42, 43]. This rule adds biodegradable bio-based mulch film to the National List. It is interesting to note that biodegradable bio-based mulch film is defined as a synthetic mulch film that meets the following criteria:

(1) Meets the compostability standards of ASTM D6400 or D6868, or of other equivalent international standards, i.e., EN 13432, EN 14995 or ISO 17088;
(2) Demonstrates at least 90% biodegradation absolute or relative to microcrystalline cellulose in less than 2 years in soil, according to ISO17556 or ASTM D5988 testing methods; and
(3) Must be bio-based with content determined using the ATM D6866 testing method.

The first requirement is meant to verify that the product satisfies the ecotoxicity requirements of the compostability tests. The second requirement is about biodegradation in soil and defines the pass levels. The last requirement refers to the bio-based content, as this is a specific prerequisite for products and materials meant for organic farming [44].

6.2.5 Standards for Testing Degradation of Mechanical Properties of Mulching Films During Their Useful Lifetime

Apart from the end-of-life requirements for total biodegradation in soil, preferably before the next cultivation season, without ecotoxicity effects, bio-based mulching films should also meet a set of minimum design requirements, including: adequate strength and elongation at break for mechanical installation and good mechanical properties with regard to ageing during the useful lifetime of the film. Specifically for bio-based low-tunnel films, not yet commercially available, the design requirements also include the adequate mechanical behavior of these films to resist various loads and load combinations (wind, hail, snow loads, etc.) [8].

The performance of bio-based biodegradable mulching and low-tunnel films over their useful lifetime may be comparable to that of conventional films in terms of tensile strength while thin bio-based films may exhibit a very low elongation at break values within the first week of their exposure. The elongation at break value is a sensitive property for thin bio-based soil-biodegradable mulching films as it degrades rather quickly when the thin films (less than 15 μm) are exposed to field conditions as compared to conventional films [8]. However, tensile strength remains steady and the whole mechanical behavior of bio-based films appears to be satisfactory provided that installation ensures the holes opened for transplantation do not result in an initial tear that might lead to subsequent tear propagation.

Thin bio-based biodegradable mulching films may also be susceptible to penetration by some weed species (e.g., *Cyperus rotundus* [45]). For this reason, resistance to penetration should be included in the mechanical properties to be tested along with testing of the tensile and tear resistance properties of bio-based mulching films. Barrier properties and radiometric properties are also important functional properties for bio-based mulching films as they affect the soil water content and temperature and plant protection.

6.3   Standards Under Preparation

Plastics and products claimed to biodegrade in soil are already available on the European market. Unfortunately, no standard specifications are available at European or International level. The reference does not necessarily have to be a European or International standard, but this would clearly be desirable in a globalized economy. Any degradability claim is potentially misleading if not referred to a testing scheme, specific test methods and specific performance requirements. This is because any organic material can be shown to be "biodegradable" in soil in the very long term (e.g., in hundreds of years).

Even materials which show a very slow rate of biodegradation or abiotic degradation can be considered as "biodegradable," if the time scale of the degradation process is of no interest. This is clearly not the case: the timescale is of great importance if biodegradation is to be used to solve the waste problem. The impact on our society of polymers with degradation times of decades and decades is different from the impact of fast biodegrading polymers for which biodegradation in soil can be confirmed in the laboratory by means of standard testing methods. The key factor in waste management is the rate of waste production that must be counterbalanced by a removal rate of similar magnitude. Rates of addition and biodegradation must be similar, otherwise an environmental buildup of plastic waste will occur.

It is self-evident that the commercialization of plastics claimed to be degradable in soil (whether biodegradable, photodegradable or another degradation characterization) without a proper testing framework is not a desired situation because it can be the premise for misleading claims to consumers and to all stakeholders who cannot translate the term "degradable" into sensible and unequivocal information. The term "biodegradable" by itself is no more informative than the adjective "tasteful" used to advertise food products [46].

In order to fill this gap, several projects have been started in recent years both at European (CEN) and North America (ASTM) level. In spite of long and deep discussions, the final outcomes have been quite limited up to now. This is because different products based on different technologies and with different environmental performance have been introduced onto the market while new products have been developed and a common denominator seems difficult to achieve.

6.3.1 Different Technologies

The technologies used to make plastic products degradable in soil are basically two.

1. The first class is based on soil biodegradable polymers, mainly polyesters and natural polymers. The materials in the first class are inherently biodegradable in soil. In other terms, the virgin polymers and plastics belonging to this class can be shown to be biodegradable already at the stage of plastic granulates and no environmental activation is needed. Biodegradability is an intrinsic property of the plastic and implies that biodegradation will start immediately after exposure to soil. This characteristic differentiates this class from the other, therefore it is relevant for standardization purposes and a proper terminology is needed to characterize this class of polymers/plastics.
2. The second class is based on traditional nonbiodegradable polymers supplemented with additives. It covers plastics that are not inherently degradable as virgin plastics but that are purported to become degradable when an additive is added during conversion. The materials that belong to the second class are mainly made starting from traditional polyethylene. The class is subdivided into two subclasses on the basis of the additive's mechanism of action.

    2.1 Additives that promote photo/thermodegradation. Polyethylene is made photo/thermodegradable under the action of the added catalysts that promote oxidation triggered by light and heat. The additives that make polyethylene degradable are known commercially as "pro-oxidants" and the polyethylene that contains pro-oxidants as "oxo-biodegradable," because the oxidized fragments that originate from the degradation are presumed to be biodegradable in the long term. This technology was originally developed in the 70s by Scott [47].
    2.2 Additives that promote enzymatic activity. Fragmentation and biodegradation of polyethylene in the long term is claimed because the additives are themselves biodegradable and "attract" microbes that presumably could also oxidize the polyethylene. This technology was originally developed in the 70s by Griffin [48].

It is not the intention of this chapter to describe the different products and the lively discussion about the real effectiveness of some of the technology involved. The interested reader can refer to review reports about the two subclasses [49, 50] and to the review paper [51] for biodegradable agricultural films. The long-term degradation behavior of oxo-degradable mulching films in soil was experimentally investigated with the burial of experimental oxo-degradable mulching films in the soil under real field conditions for approximately eight years. The degradation behavior of these films, used with melon cultivation before burial or artificially degraded through accelerated UV radiation before burial, is presented in [52, 53], respectively.

It should be remarked that standards about the environmental fitness of products (i.e., polymers and plastics) should be technology neutral. The focal point is the environment; therefore the starting point must be the environment and not the product or the technology. Test conditions, biodegradation times and acceptability levels must be correlated with the environment and not with the characteristics of any specific material. It is the material that must meet the biodegradability criteria under specific natural environmental conditions and not the opposite, and a testing scheme should aim to discriminate between products that satisfy some minimal environmental requirements and products that do not. Therefore, any material is suitable as long as it satisfies the environmental requirements.

6.3.2   Discussion on Soil Biodegradation at CEN TC 249 WG9

A long discussion took place at the CEN Technical Committee TC 249 (*Plastics*) level in the Working Group WG9 (*Characterization of degradability*) on how to structure a standard specification for biodegradation in soil. The group was not able to reach a consensus but it was decided to keep track of this discussion with a Technical Report CEN/TR 15822 [54].

The points of agreement were the following:

(1) Soil cannot be considered as a dumping location for plastic particles, even if they are proven to be inert.
(2) There should be no long-term accumulation of fragments, no release of harmful degradation species or additives and no negative effects on crops or on soil fertility. This implies that:

- plastics added to soil should convert to CO_2 and biomass sufficiently rapidly to ensure that they do not show long-term accumulation in the soil.
- neither the plastics themselves nor their degradation products and residues should be toxic to microorganisms, macroorganisms (e.g., worms), plants or the animals that consume them. Furthermore, they should neither negatively influence the germination of seeds or the yield of crops, nor have

unacceptable environmental impacts on air, water, and soil. Standard eco-toxicity tests defined in existing compost standards are considered adequate, using relatively high initial concentrations of the material under study to simulate repeated applications.

(3) The evaluation of biodegradability of plastics, as materials, should be based on conditions that are as representative as possible of real field conditions (e.g., soil burial). The requirements have to meet market quality needs and social acceptability criteria.

(4) A standard test for disintegration would be helpful in order to determine the time for the substantial disappearance of the plastic product, once inherent biodegradability has been tested using other test methods.

The major point of disagreement was the issue of "pretreatment"—exposure of samples to light/heat realistically representing the field conditions, before testing biodegradation. It was agreed that different tests will inevitably be required for materials intended for different applications where pre-biodegradation lifetimes are so variable.

The ideal testing scheme resulting from the WG9 discussion was based on 3 possible pretreatment routes: (1) a pretreatment based on light/heat to simulate exposure to sunlight; (2) a pretreatment based only on heat to simulate burial in sandy top soil, where temperature can significantly increase through sun irradiation; (3) no pretreatment, to simulate direct soil burial without any direct or indirect exposure to sunlight.

Irradiation parameters (intensity, duration, etc.) should be well defined in order to avoid unrealistic overtreatment and should consider the geographical area of application. Such testing should be carried out using instruments designed to simulate solar radiation, with control of temperature and humidity.

The biodegradation of plastic samples after the three exposure pathways should be measured by applying the test method described in ISO 17556 or similar test methods operating in the so-called mesophilic temperature range, and showing a 90% biodegradation level relative to an appropriate reference material in less than 2 years. The WG9 also defined what an "appropriate reference material" is: a homogeneous material which fully biodegrades during the test period chosen for the particular application. In practice this means that mineralization at the plateau level shall be equal to or higher than 60% in less than 2 years.

The CEN/TR 15822 [54] also suggests that if the test material fails to reach relative biodegradation of 90% in two years but reaches 60% (absolute) in 1 year, this could be considered as proof that the material is potentially biodegradable to be confirmed by additional substantiation by performing an accelerated mineralization test method (e.g., EN ISO 14855 [55] performed at 58 °C), and reaching 90% (relative to an appropriate reference material) in 6 months. The application of an accelerated test method is contentious though, as for many experts this approach is very different from the soil conditions.

6.3.3   Draft Standard Specification for Biodegradable Mulching Films

Despite all this, progress has been made at the CEN level by finally succeeding to develop a draft of standard specifications for Biodegradable mulch films for use in agriculture and horticulture—Requirements and test methods [56]. The draft document prEN 17033:2016 [56] prepared by the Technical Committee CEN/TC 249 "Plastics," was submitted to the CEN Enquiry in August 2016. The main distinct section of this draft that differs from the standard specification for conventional agricultural films EN 13655, 2002 [57] concerns the requirements for materials, testing schemes and evaluation criteria for biodegradation and ecotoxicity. Special emphasis is also placed on defining the service life on soil of biodegradable mulch films. Other sections concern dimensional, mechanical, and optical properties of the films and some provisions for delivery, storage, marking, installation, etc. The mechanical and optical properties requirements of prEN 17033 [56] for unexposed biodegradable films are analogous to those defined by EN 13655 [57] for conventional black mulching films with thickness less than 30 μm, considering that the draft document prEN 17033 [56] refers to three thickness classes: <10 μm, 10–15 μm and >15 μm.

6.3.4   Discussion on Soil Biodegradation at ASTM D20.96

During a 3-year long discussion, the ASTM D20.96 subcommittee attempted to establish a standard specification on Biodegradable plastics in soil environment. Similarly to what had happened at CEN level, this attempt encountered quite a lot of opposition. The discussion is still ongoing. The main arguments are shown below because they are of general interest for the community interested in standard developments.

6.3.4.1   Scope: About Plastic Material or Final Products?

The first argument was about the scope, whether it should cover the basic plastic material or the final product. Specifications that cover the characteristics of plastic do not consider any particular application. Any discussion about the specific requirements of any particular application is postponed to other standards covering the specific products of interest.

The implication of such an early decision about the desired scope is manifold.

If the standard covers only the biodegradability of the plastic material (i.e., not the final product) then "Environmental Degradation," i.e., the abiotic degradation that occurs when the plastic product is laid on soil and exposed to atmospheric factors such as sunlight, rainfall, humidity, etc., is not comprised. On the other hand, if the standard covers the biodegradability of the product, for instance of mulch films, then "Environmental Degradation" is an appropriate phase of the lifetime of the product that should be taken into consideration.

The subcommittee was not able to clarify this point and several draft standards were discussed and balloted. The scope of some of these also comprised products made from biodegradable plastics while others were only focused on the basic material (named *plastics that are innately biodegradable in soil* or *virgin Plastics that biodegrade in soil*). Some experts expected the standard to consider the final products but to omit Environmental degradation, while others expected the standard to consider the material only, but to cover Environmental degradation as well. It is easy to understand that these requirements are incompatible and contradictory. If the standard is meant to consider the characteristics of the virgin material then no environmental degradation is to be taken into consideration as this will depend on the specific application (whether it is a mulch film or other products). On the other hand, if the standard is meant to cover the final applications as well, then environmental degradation cannot be excluded in an arbitrary way.

6.3.4.2   Classification

An attempt was also made to classify the products into two classes in order to accommodate the different technologies. One class was defined as "Plastics biodegradable in Soil Environment" covering materials inherently biodegradable in soil. This class was considered as suitable for any application where the plastic products are buried in soil after or during the service life. The other class, "Plastics biodegradable in Soil Environment after environmental degradation," was thought to describe those materials that are expected to be biodegradable in soil only after environmental degradation has occurred in suitable times and conditions. The latter class was devised to consider materials suitable for making products that are wholly exposed to atmospheric factors so that they acquire sufficient biodegradability in soil burial after exposure. This class was considered as unsuitable for making products that are in part exposed and in part directly buried in soil or for those products that are only briefly exposed to sun irradiation.

6.3.4.3   Environmental Degradation Must Be Linked to a Specific Geographical Zone

If plastic products become biodegradable after "environmental degradation," then it is very relevant to define the exposure conditions. Needless to say, the environmental degradation that is experienced in a desert is very different to that experienced in a field in the middle of Europe. Therefore, the environmental degradation that can be obtained in a specific geographical zone can be different in another zone and results could differ. It is relevant to well define the exposure condition applied in order to make a sound correlation with the result that can be obtained in real application. During the discussion at ASTM level it was proposed to consider the typical environmental conditions found in the Temperate Zone as defined by the area between the cold zone and the subtropical zone (latitude from 40° to 60°) with an average temperature between 0 and 20 °C (minimum temperatures: −40 °C; maximum temperatures: +40 °C); day length from 4 to 16 h; precipitation from 300 to 2000 mm (average 800 mm); annual global irradiance between 900 and 1700 kW h/m². This was rejected as many experts expressed their desire to test any possible environment. This is clearly an approach that makes standardization difficult.

6.3.4.4 Terminology

A vivid discussion took place at ASTM level on the terminology to be used to describe materials that are intrinsically biodegradable in soil. It was proposed to describe this class as "plastics that are innately biodegradable in soil," but this was rejected. It was then proposed to designate this class as "virgin plastics that biodegrade in soil." This was also rejected in spite of the fact that the term "virgin" is actually a defined term in ASTM: "virgin plastic—a plastic material in the form of pellets, granules, powder, floc or liquid that has not been subjected to use or processing other than that required for its initial manufacture" (ASTM 883 [58]). Materials that are biodegradable in soil even before conversion into a specific product must be designated, because this is relevant information in order to discriminate them from products that are assumed to become degradable only after conversion together with specific additives.

6.3.4.5 Biodegradation Rate

The proposed biodegradation threshold under discussion is 90% in 2 years, to be shown by applying either ISO 17556 or ASTM D5988, in line with the specifications of the USDA-AMS National Organic Program (NOP) for mulch films allowed for organic crop production [44]. The test duration of 2 years is contentious and criticized for being too short by some ASTM members and too long by other members.

6.4 Functional Barriers with Respect to Standards and Labeling

A combination of open issues, contradictory information and terminology about bio-based materials used in the Agrofood sector has created significant confusion in the sector for both stakeholders and the public.

6.4.1 Communication

Characteristic examples described in the literature concern terminology issues that lead to a confusing perception of bioproduct labels for the consumer (e.g., bio-based versus biodegradable, versus degradable, versus sustainable, versus recyclable). Key communication issues include:

- Bio-based films are not necessarily biodegradable and if they are biodegradable they are not necessarily subjected to biodegradation in any environment and conditions, such as in soil. Bio-based mulching films that are biodegradable in soil should be labeled as such based on the relevant standard testing methods and specifications.
- The introduction of fragmentable or oxo-degradable polyethylene mulching films in agricultural practice has created real confusion for farmers, plastic converters and the market. Polyethylene mulching films use specialized pro-oxidant additives that accelerate the breakdown of polyethylene to very small fragments under the action of UV radiation and temperature. Some authors claim that the oxidation products of polyolefins may be biodegradable [59–61]. The photochemical and thermal degradation of these products under artificial laboratory conditions is presented in several research works in the technical literature [62–65] and is supported by the relevant industry. However, the ultimate biodegradability of these materials is in question [66], as contradictory results have also been reported in the literature while biodegradability of such materials in real soil conditions is strongly disputed [67–69]. In practical terms, farmers think that these relatively cheap films, as compared to bio-based, soil-biodegradable mulching films, are biodegradable and even confuse them with bio-based materials as they are not aware of the associated controversy [51].
- The confusion about terminology described above for the bio-based materials used in agriculture may be associated with additional vague claims such as "sustainable," "environmentally friendly," "eco-friendly," etc., which create real confusion for the industry, farmers and the public.

6.4.2 Confusion in Communication with Standards

A major confusion in the market concerns the misunderstanding or misuse of existing standards. Examples of problematic claims in the bio-based plastics market are presented in [70]. According to [68], this is a common misconception: Standard Test Methods are used to determine the *rate* of biodegradation and the *degree* of biodegradation of a material, such as a mulching film; but they do not determine whether a mulching film meets any particular requirement, and have nothing to do with certification or labeling. Conversely, standard specifications for biodegradability in a specific environment (e.g., soil, compost or fresh water) define a set of requirements that all have to be satisfied by a material, product, system, or service, in order to be designated the particular label. In addition, standards do not use phrases such as "*safe for the environment*" as used by some industries in association with a Standard Test Method [68].

Bio-based materials, like bio-based mulching films that are biodegradable in soil, should indicate that they are certified by a third-party agency which confirmed the tests, rather than implying that meeting the test requirements (without even reference to test method and specifications) is enough for certification [68]. The standard tests should be explicitly mentioned (e.g., test method for biodegradability of mulching film in soil).

With regard to the determination of bio-based content, the ASTM D5988 standard is only a test. It does not make sense to indicate "meeting the requirements" for D5988 because a test method does not specify a "required" measurement to pass or fail [68].

In some cases standards are confused with certifications. Thus, ASTM D5988 is not a "certification" (standards are not certifications, because a certification is the result of having the results of a test determined by a third party and includes continuous surveillance of the products on the market) and it is not even a specification; it is only a test method.

Further confusion, particularly for biodegradable mulching films, arises from claims of biodegradability in soil without clear reference to a standard. Because of these kinds of problems the Federal Trade Commission in the United States has recently been grappling with the issue of environmental claims that companies make about their products [68] and similar actions are in progress in the EU.

Many companies claim that their products "conform to" an ISO or ASTM standard test method for biodegradation, or "pass" the ISO or ASTM standard test method for biodegradability, or "can be defined as biodegradable" based on an ISO or ASTM standard. All of these claims are incorrect. ISO and ASTM have defined several test methods for biodegradation, using different methods and under different conditions. There are tests for many different types of biodegradation under many different conditions. The results of these tests are biodegradation levels that can range from 0% (no biodegradation) to 100% (total biodegradation) over the testing period and under the testing conditions. None of the tests for biodegradation produce a "pass/fail" result [68].

It should be noted that the existing specifications, like the D6400 specification, were never meant to be universal specifications for whether something is "biodegradable" or "compostable." The test methods and specifications refer to very specific biodegradation or composting environments and specific test materials. This explains why more and more standard test methods and specifications are under development or planned to be developed.

6.5   Conclusions

The main international standard test methods for biodegradation of plastics in soil are: ISO 17556, ASTM D5988, NF U52-001 and UNI 11462. These standard testing methods determine the rate of biodegradation of plastics in soil media under normalized conditions. The normalized conditions defined by these standard testing methods differ in several aspects related to (a) soil medium (natural soils collected from specified locations, a laboratory mixture of soils, a mixture of natural soil and mature compost or a "standard soil"); (b) test sample (intact, sample cut into pieces or pulverized); (c) soil pH (natural pH or adjusted); (d) C/N ratio (natural C/N ratio or adjusted to 10:1–20:1 with respect to the mass of organic carbon contained in the sample or to 40:1 for sample organic carbon to soil N) and (e) water content (adjusted with respect to water holding capacity).

The standard testing procedures may enhance the conditions for biodegradation of the tested polymers. This may lead to an optimal controlled biodegradation process for achieving repeatability of the testing method results. However, these results may not be representative of the actual biodegradation rates of the same tested materials/products experienced under natural conditions. Although results may indicate that the tested plastic material will biodegrade under the test conditions at a certain rate, the standards caution that the laboratory testing results may not be extrapolated to real soil environments since soil properties, temperatures and water content vary widely and could be very different from the soil media used in the laboratory testing while they also vary with time.

Besides the biodegradation test methods, pass levels and a time frame also need to be defined in order to determine whether a bio-based product will biodegrade sufficiently under soil conditions. There is currently no European or international specification that defines criteria for biodegradation of bio-based products in soil. The time frame and pass levels for biodegradation of biodegradable materials used in agriculture and horticulture are only defined in standard specifications NF U52-001 and UNI 11462, together with criteria for environmental safety. However, the evaluation of the biodegradability in soil is not obligatory in the French specification.

According to De Wilde [71], as far as acceptance of biodegradable plastics in soil is concerned, two main categories of testing requirements are defined:

(1) Biodegradation: 90% with duration of 2 years for mulching films in function of the application
(2) Soil quality: no ecotoxicological effects, material may not contain heavy metals. These proposed specifications for biodegradation in soil are analogous, to some extent, to those of the French Standard NF U52-001 (apart from the optional use of the soil medium by NF U52-001), extended over a longer period to determine long-term effects. As mentioned by Briassoulis and Dejean [20], analogous provision has been adopted by the SP Technical Research Institute of Sweden concerning the requirements and associated test methods to certify polymeric materials and products: the ultimate aerobic biodegradability in soil

(by measuring the oxygen demand in a respirometer or the amount of carbon dioxide evolved) is determined by the requirement of biodegradation $\geq 90\%$ within 24 months. The time frame ($\geq 60\%$ in 12 months) and biodegradation rate requirements of NF U52-001 do not guarantee long-term biodegradation in soil as suggested by De Wilde B. [69].

Acknowledgments Part of this work is based on the relevant state-of-the-art review carried out within the framework of the KBBPPS project supported by the European Commission through the FP7 Programme (FP7-KBBE-2013-312060). Special thanks are due to Nike Mortier and Bruno De Wilde (OWS) and to Antonis Mistriotis (AUA), for contributing to the corresponding work of KBBPPS [19, 21].

References

1. Hopewell J, Dvorak R, Kosior E (2009) Plastics recycling: challenges and opportunities. Philos Trans R Soc London [Biol] 364:2115–2126
2. Plastics Europe (2013) Plastics—the Facts 2013, an analysis of European latest plastics production, demand and waste data. Available via; http://www.plasticseurope.org/documents/document/20131014095824-final_plastics_the_facts_2013_published_october2013.pdf. Accessed 22 Oct 2015
3. European Bioplastics. Available at: http://www.europeanbioplastics.org/. Accessed 15 Sep 2013
4. Briassoulis D, Hiskakis M, Scarascia G, Picuno P, Delgado C, Dejean C (2010) Labeling scheme for agricultural plastic wastes in Europe. Qual Assur Saf Crop 2(2):93–104
5. Shogren RL, Hochmuth RC (2004) Field evaluation of watermelon grown on paper-polymerized vegetable oil mulches. HortScience 39:1588–1591
6. Briassoulis D, Babou E, Hiskakis M, Scarascia G, Picuno P, Guarde D, Dejean C (2013) Review, mapping and analysis of the agricultural plastic waste generation and consolidation in Europe. Waste Manage Res 31:1262–1278
7. Washington State University Extension (2013) Using biodegradable plastics as agricultural mulches, fact sheet FS103E, Washington State University, WSU Extension. Available via; http://pubs.wsu.edu. Accessed 28 Apr 2014
8. Briassoulis D (2007) Analysis of the mechanical and degradation performance of optimised agricultural biodegradable films. Polym Deg Stab 92(6):1115–1132
9. Degli Innocenti F (2005) Biodegradation behaviour of polymers in the soil. In: Bastioli C (ed) Handbook of biodegradable polymers. RAPRA Technology Limited Shawbury, Shrewsbury, pp 57–102
10. Degli Innocenti F (2011) The role of standards for biodegradable plastics. Bioplastics Mag 4:36–38
11. De Wilde B (2014) International and national norms on biodegradability and certification procedures. In: Bastioli C (ed) Handbook of biodegradable polymers, 2nd edn. Smithers Rapra Shawbury, Shrewsbury, pp 139–174
12. ISO/IEC Guide 2 (1996) Standardization and related activities—general vocabulary, definition 1.1. International Organization for Standardization, Switzerland
13. EN 45020 (2007) Standardisation and linked activities—general Vocabulary
14. Degli Innocenti F (2003) Biodegradability and compostability—the international norms. In: Chiellini E, Solaro R (eds) Biodegradable polymers. Plastics Kluwer Academic Plenum Publishers, New York, pp 33–45

15. ISO 17556 (2012) Plastics—determination of the ultimate aerobic biodegradability of plastic materials in soil by measuring the oxygen demand in a respirometer or the amount of carbon dioxide evolved. International Organization for Standardization, Switzerland
16. ASTM D 5988 (2012) Standard test method for determining aerobic biodegradation in soil of plastic materials. ASTM International, USA
17. EN ISO 17556 (2012) Plastics—determination of the ultimate aerobic biodegradability of plastic materials in soil by measuring the oxygen demand in a respirometer or the amount of carbon dioxide evolved. International Organization for Standardization, Switzerland
18. ISO 11266 (1994) Soil quality—guidance on laboratory testing for biodegradation of organic chemicals in soil under aerobic conditions. International Organization for Standardization, Switzerland
19. KBBPPS Project (2013) Deliverable N° 6.2: draft biodegradability standard, FP7 Programme (FP7-KBBE-2013- 312060). Available via; http://www.biobasedeconomy.eu/research/kbbpps/. Accessed 21 July 2016
20. Briassoulis D, Mistriotis A, Mortier N, De Wilde B (2014) Standard testing methods & specifications for biodegradation of bio-based materials in soil—a comparative analysis. In: Proceedings international conference of agricultural engineering, Ref: C0668, Zurich, 06–10.07.2014. Available via; http://www.geyseco.es/geystiona/adjs/comunicaciones/304/C06680001.pdf. Accessed 21 July 2016
21. Briassoulis D, Dejean C (2010) Critical review of norms and standards for biodegradable agricultural plastics, part I. Biodegradation in soil. J Polym Environ 18(3):384–400
22. De Wilde B, Mortier N, Verstichel S, Briassoulis D, Babou M, Mistriotis A, Hiskakis M (2013) Report on current relevant biodegradation and ecotoxicity standards, Deliverable D1 of project: knowledge based bio-based products (KBBPPS; FP7-KBBE-2013-312060), Ghent. Available via; http://www.kbbpps.eu. Accessed 31 Jan 2013
23. ISO 9408 (1999) Water quality—evaluation of ultimate aerobic biodegradability of organic compounds in aqueous medium by determination of oxygen demand in a closed respirometer. International Organization for Standardization, Switzerland
24. NF U52-001 (2005) Biodegradable materials for use in agriculture and horticulture-mulching products-requirements and test methods. Association Française de Normalisation
25. UNI 11462 (2012) Plastic materials biodegradable in soil—types, requirements and test methods. Italian Organization for Standardization (UNI)
26. ISO 11269-2 (2012) Soil quality—determination of the effects of pollutants on soil flora—part 2: effects of contaminated soil on the emergence and early growth of higher plants. International Organization for Standardization, Switzerland
27. FD X 31-251 (1994) Qualité du sol – Effets des polluants vis-à-vis des vers déterre (*Eisenia*) (*fetida*). Partie 1: détermination de la toxicité aigue en utilisant des substrats de sol artificiel. Statut: fascicule de doc. AFNOR 11 p
28. NF T 90-375 (1998) Qualité de l'eau - Détermination de la toxicité chronique des eaux par l'inhibition de la croissance de l'algue d'eau douce Pseudokirchneriella subcapitata (Selenastrum capriconutum). AFNOR. 13 p
29. UNI 11495 (2013) Biodegradable thermoplastic materials for use in agriculture and horticulture—mulching films—requirements and test methods. Italian Organization for Standardization (UNI)
30. UNI 10780 (1998) Compost—classification, requirements and use criteria. Italian Organization for Standardization (UNI)
31. ISO 6341 (2012) Water quality—determination of the inhibition of the mobility of Daphnia Magna Straus (Cladocera, Crustacea)—acute toxicity test. International Organization for Standardization, Switzerland
32. Belgian Royal Decree (2008) Decree specifying the norms that products should meet to be compostable or biodegradable. Official Journal of 28 October 2008, Effective in July 2009
33. OECD (1984) Test No. 207: earthworm, acute toxicity tests, OECD guidelines for the testing of chemicals (Section 2). OECD Publishing, Paris. doi:http://dx.doi.org/10.1787/9789264070042-en

34. OECD (2006) Test No. 208: Terrestrial plant test: seedling emergence and seedling growth test, OECD guidelines for the testing of chemicals (Section 2). OECD Publishing, Paris. doi: http://dx.doi.org/10.1787/9789264070066-en
35. OECD (2004) Test No. 222: Earthworm reproduction test (Eisenia fetida/Eisenia andrei), OECD guidelines for the testing of chemicals (Section 2). OECD Publishing, Paris. doi:http://dx.doi.org/10.1787/9789264070325-en
36. OECD (2010) Test No. 317: Bioaccumulation in terrestrial oligochaetes, OECD guidelines for the testing of chemicals (Section 3). OECD Publishing, Paris. doi:http://dx.doi.org/10.1787/9789264090934-en
37. ISO 11268-1 (2012) Soil quality—effects of pollutants on earthworms—part 1: determination of acute toxicity to Eisenia fetida/Eisenia Andrei. International Organization for Standardization, Switzerland
38. ISO 22030 (2005) Soil quality—biological methods—chronic toxicity in higher plants. International Organization for Standardization, Switzerland
39. ASTM E1676 (2012) Standard guide for conducting laboratory soil toxicity or bioaccumulation tests with the Lumbricid Earthworm Eisenia Fetida and the Enchytraeid Potworm Enchytraeus albidus. ASTM International, USA
40. European Commission (DG Environment) (2011) Plastic waste in the environment, Final Report April 2011. Available via; http://ec.europa.eu/environment/waste/studies/pdf/plastics.pdf. Accessed 21 July 2016
41. AIB-VINÇOTTE International S.A./N.V, Member of the Group AIB-VINÇOTTE. Available via; http://www.vincotte-certification.com/en/home/. Accessed 21 July 2016
42. Agricultural Marketing Service (2016) United States Department of Agriculture. Electronic Code of Federal Regulations, PART 205—National Organic Program. Available via; https://www.ams.usda.gov/rules-regulations/organic/national-list. Accessed 21 July 2016
43. Corbin et al (2014). Current and future prospects for biodegradable plastic Mulch in certified organic production systems—eXtension, Organic Agriculture. Available via; http://www.extension.org/...7951/current-and-future-prospects-for-biodegradable-plastic-mulch-in-certified-organic-production-systems. Accessed 28 Apr 2014
44. Agricultural Marketing Service (2013) USDA. National Organic Program; Proposed: amendments to the national list of allowed and prohibited substances (crops and processing), 7 CFR Part 205, Document Number AMS–NOP–13–0011, NOP–13–01PR. Available via; http://www.gpo.gov/fdsys/pkg/FR-2013-08-22/pdf/2013-20476.pdf. Accessed 21 Jul 2016
45. Anzalone A, Cirujeda A, Aibar J, Pardo G, Zaragoza C (2010) Effect of biodegradable mulch materials on weed control in processing tomatoes. Weed Technol 24(3):369–377
46. Society of the Plastics Industry Bioplastics Council (2010) Position paper on oxo-biodegradables and other degradable additives. Available via; http://spi.files.cms-plus.com/about/BPC/SPI%20Bioplastic%20Council%20Bioplastics%20Position%20Paper%20on%20OXO-Biodegradable%20Plastic-FINAL.pdf. Accessed 22 Oct 2015
47. Scott G, Wiles DM (2001) Programmed-life plastics from polyolefins: a new look at sustainability. Biomacromolecules 2001(2):615–622
48. Griffin GJL (1974) Biodegradable fillers in thermoplastics. Adv Chem 134:159–170
49. Deconinck S, de Wilde B (2013) Benefits and challenges of bio- and oxo-degradable plastics, a comparative literature study, OWS, DSL-1, Aug-09-2013. Available via; http://ows.be/wp-content/uploads/2013/10/Executive-summary1.pdf. Accessed 22 Oct 2015
50. Deconinck S, de Wilde B (2014) Review of information on enzyme-mediated degradable plastics, OWS, EUBR-2, May-10-2014. Available via; http://ows.be/wp-content/uploads/2014/08/Report_Rev01.pdf. Accessed 22 Oct 2015
51. Kyrikou I, Briassoulis D (2007) Biodegradation of agricultural plastic films—a critical review. J Polym Environ 15(2):125–150
52. Briassoulis D, Babou E, Hiskakis M, Kyrikou I (2015) Analysis of long term degradation behaviour of polyethylene mulching films with pro-oxidants under real cultivation and soil burial conditions. Environ Sci Pollut Res 22:2584–2598. doi:10.1007/s11356-014-3464-9

53. Briassoulis D, Babou E, Hiskakis M, Kyrikou I (2015) Degradation in soil behaviour of artificially aged polyethylene films with pro-oxidants. J Appl Polym Sci 132(30):1–20. doi:10.1002/app.42289

54. CEN/TR 15822 (2009) Plastics—biodegradable plastics in or on soil—recovery, disposal and related environmental issues, Technical Report, Technical Committee CEN/TC 249, European Committee for Standardization

55. ISO 14855-1 (2012) Determination of the ultimate aerobic biodegradability of plastic materials under controlled composting conditions—method by analysis of evolved carbon dioxide—part 1: general method. International Organization for Standardization, Switzerland

56. prEN 17033 (2016) (E) Biodegradable mulch films for use in agriculture and horticulture—requirements and test methods European Committee for Standardization, Brussels, Belgium Edition (This document is currently submitted to the CEN Enquiry)

57. EN 13655 (2002) Plastics—mulching thermoplastic films for use in agriculture and horticulture, European standard. European Committee for Standardization, Brussels, Belgium

58. ASTM D883 (2012) Standard terminology relating to plastics, ASTM International, West Conshohocken, PA, 2012. www.astm.org

59. Wiles DM, Scott G (2006) Polyolefins with controlled environmental degradability. Polym Degrad Stab 91(7):1581–1592

60. Jakubowicz I (2003) Evaluation of degradability of biodegradable polyethylene (PE). Polym Degrad Stab 80(1):39–43

61. Scott G, Wiles DM (2001) Programmed-life plastics from polyolefins: a new look at sustainability. Biomacromolecules 2(3):615–622

62. Karlsson S, Albertsson AC (1998) Biodegradable polymers and environmental interaction. Polym Eng Sci 38:1251–1253

63. Matsumaga M, Whitney PJ (2000) Surface changes brought about by corona discharge treatment of polyethylene film and the effect on subsequent microbial colonisation. Polym Deg Stab 70:325–332

64. Broska R, Rychly J (2001) Double stage oxidation of polyethylene as measured by chemiluminescence. Polym Deg Stab 72:271–278

65. Bonhomme S, Cuer A, Delort AM, Lemaire J, Sancelme M, Scott G (2003) Environmental biodegradation of polyethylene. Polym Deg Stab 81(3):441–452

66. Singh B, Sharma N (2008) Mechanistic implications of plastic degradation. Polym Degrad Stab 93(3):561–584

67. Feuilloley P (2004) Ce plastique faussement biodegradable. La Recherche 374:52–54

68. Fritz J (2003) Strategies for detecting ecotoxicologic effect of biodegradable polymers in agricultural application. Macromol Symp 197:397–409

69. Thompson RC, Olsen Y, Mitchell RP et al (2004) Lost at sea: where is all the plastic? Science 304(5672):838. doi:10.1126/science.1094559

70. Stevens G (2010) Bioplastic standards 101, green plastics. Available via; http://green-plastics.net/news/45-science. Access 28 Apr 2014

71. De Wilde B (2002) Standardization activities related to measuring biodegradability of plastics in soil and marine conditions. Paper presented at the Congress Industrial Applications of Bio-Plastics 2002, York, UK, 3–5 Feb 2002

Chapter 7
Life Cycle and Environmental Cycle Assessment of Biodegradable Plastics for Agriculture

Francesco Razza and Alessandro K. Cerutti

Abstract The study of the life cycle of products for the quantification of their environmental impacts, in each of their production and utilization stages, is a well-established and scientifically recognized methodology. This approach is known as Life Cycle Assessment (LCA) and it is the base for several product and service certifications. This chapter focuses on the strengths and weakness of the LCA approach to biodegradable plastic in agriculture, thorough the description of main issues emerged from an extensive literature search and key case studies. In particular, studies which apply LCA on biodegradable mulching films and nursery pots are presented and discussed. Results of the study are somehow controversial. Despite the fact that LCA is the most systematic way to understand the interrelation between a product and the environment (including biodegradable products), some specific issues, related to the own nature of biodegradable products, require a more detailed way to be properly addressed. In particular crucial issues are related to the modelling of realistic waste management scenarios, the development of more appropriate impact categories (e.g. effects of littering) and the assessment of biorefineries which represent the only non-fossil carbon source for bio-based plastic materials.

Keywords Sustainable agriculture · Environmental impact assessment · Nursery systems · Mulch film · Waste management · White pollution · Life cycle assessment · Carbon footprint

F. Razza (✉)
Novamont, Terni, Italy
e-mail: Francesco.Razza@novamont.com

A.K. Cerutti
Department of Agriculture, Forestry Food Science, University of Turin, Turin, Italy

© Springer-Verlag GmbH Germany 2017
M. Malinconico (ed.), *Soil Degradable Bioplastics for a Sustainable Modern Agriculture*, Green Chemistry and Sustainable Technology,
DOI 10.1007/978-3-662-54130-2_7

7.1 Elements of Life Cycle Assessment

The term Life Cycle Assessment (LCA) refers to a technique to assess the environmental aspects and potential impacts associated with a product, process, or service. The LCA protocol includes four phases:

- Goal and scope definition
- Compiling an inventory of relevant energy and material inputs and environmental releases
- Evaluating the potential environmental impacts associated with identified inputs and releases
- Interpreting the results to help you make a more informed decision.

A short summary of these phases is provided below:

(I)   Goal and scope definition deals with the clear and unambiguous formulation of the research question and the intended application of the answer that the LCA study is supposed to provide [1]. Important elements of this phase are the choice of the functional unit, the definition of the system boundaries and reference flows, the description of the methodological approach (e.g. attributional or consequential). Once the goal of an LCA study is defined, the initial scope settings are derived that define the requirements of the subsequent work. Anyhow, during the execution of the analysis, more information becomes available (e.g. hotspots, relevant processes/materials involved in the product system etc.) at point that, almost always, scope settings need to be refined or sometimes revised [2]. For these reasons LCA is considered an iterative process.

(II)   The inventory analysis is concerned with the collection of quantitative data on inputs (resources and intermediate products) and outputs (emissions, wastes) for each unit process that compose the product systems. The final outcome of this phase is the inventory table of the whole analyzed product system.

(III) Within the evaluation phase the inventory data on inputs and outputs are translated into indicators about the product system's potential impacts on the environment, on human health, and on the availability of natural resources.

(IV)   Interpretation is the phase where the results of inventory and evaluating phases are interpreted according to the goal of the study and where sensitivity and uncertainty analysis are performed to qualify the results and the conclusions.

A deeply coverage of the LCA methodology is provided in the ILCD handbook series [2], whereas a comprehensive scientific treatise of its computational structure is provided in [1].

7.2 LCA Overview: From Early Studies to Nowadays

The first, rare, Life Cycle Assessment (LCA) studies date back to the early '70s, when large multinationals started to investigate the life cycle of its products in order to figure out where in the production process could save resources, primarily for economic reasons [3]. In such studies, results were reported as resource and emission profiles, but no quantitative assessment of the associated impacts on environment was performed [4].

In that period, also as a result of the early environmental campaigns, industrial enterprises realized that to be efficient in the use of resources not only led to cost savings, but also to a lesser impact on ecosystems. In the following years the LCA approach was developed and formalized to be used in the industrial sector, but it was just in the mid-90s that its validity was recognized also for the agricultural sector [5].

Nowadays, the study of the life cycle of products for the quantification of their environmental impacts, in each of their production and utilization stages, is a well-established and scientifically recognized methodology whose principles and methodological framework have been drawn in several certification schemes focused on environmental profile of product and services like ISO 14064:2013 well known as Carbon Footprint (CF), Environmental Product Declarations (EPD), Product Environmental Footprint (PEF), etc. Life cycle assessment methodology has been standardized by the International Standardisation Organisation (ISO) in the ISO 14040 series (1997–2003), and revised in 2006 (i.e. ISO 14040:2006 and ISO 14044:2006).

It is interesting to point out that up to early 2000 the difficulties in understanding the results of an LCA by non-experienced user limited its application in confined areas of research. However, over the last years, thanks to the great stir that climate change concerns have had (and still have), the situation is changed. For example the advent of the Carbon Footprint (CF) has opened the LCA results to a wider, largely non-technical audience [6]. As a result the number of studies focusing exclusively in CF or energy associated to a product or an agricultural practice is dramatically increasing, forgiving all impacts that other emissions have on ecosystems.CF, until standardization in ISO 14064:2013, was nothing more than a calculation of an LCA in which the only impact category that was present is Global Warming Potential, measured in mass of CO_2 equivalent, disregarding the principle LCA core of the completeness and the ability to evaluate the trade-off [7].

In parallel to this process of simplification of impact categories, there has been a phenomenon of juxtaposition of the idea of sustainability to LCA. In 2010 the number of scientific papers associating the idea of sustainability and LCA were three times compared to 2007 [8] and this trend is rapidly ascending until today in which virtually in every article the two concepts are associated. But is that really so? The LCA approach is used to assess the sustainability of a product or an agricultural technique? The historic route LCA approach allows to highlight how the tool has evolved with respect to its original formulation, in particular the analysis was born with the aim to quantify the efficiency (economic and environmental) of a production system, but not its sustainability. One thing is the environmental performance of a product and another is its sustainability. This decoupling is evident when comparing—as an example—systems with low use of resources, and its low rate of production, with systems in wide use of resources and its high rate of production. In the first perspective of LCA it is very often disadvantaged (especially in food systems) as, per unit of the product obtained, the agronomic inputs are lower. But often the situation reverse when instead of considering the impacts per unit of output we consider them per unit of land occupied [9] or per unit of energy applied [10] in a production system. Another example of the decoupling of LCA and sustainability is given by biodegradable materials. As will be highlighted in this chapter the LCA applied to biodegradable bioplastic products used in agriculture is very useful (i.e. hotspots identification along supply chain) and desirable (i.e. positioning of biodegradable products compared to their petroleum-based counterparts in terms of GHG emissions, resource efficiency etc.) Nevertheless, it is not rare, that the "big issues", environmentally speaking, are not properly addressed or in other terms quantitatively accounted in LCA due to methodology constrains. A well-fitting example of a "big issue" which is completely disregarded in LCA of products used in agriculture, is the "white pollution" phenomenon. It consists of the contamination of agricultural soils with non-biodegradable mulching plastic films that for different reasons, remain in the soil causing detrimental effects [11]. Such phenomenon, over the last ten years has taken impressive dimensions like in Xinjiang, China where residual plastic film mulch has become a serious issue that need to be addressed from aspects of policy, regulation and technology in all-round manner.

Summarizing, the LCA has allowed a major breakthrough in the study of agricultural systems. [12] described LCA as one of the most appropriate methods to identify, with high degree of detail, environmental hotspots, compare techniques and inform with scientific data the decision makers both at firm and political level. However, LCA application in the agri-food sector is a complex and challenging endeavour [13], it has allowed technicians to simplify agricultural systems in their mechanistic and dynamic at the same time to bring out the connections (often ignored) between the various agricultural technologies with their environmental burden. Yet still many shadows must be brought to light and quickly, because more and more of LCA results are used at the base of the construction of food policies and local development.

7.3 Overview of LCA on Biodegradable Plastics for Agriculture

The application of LCA studies on bio-based polymers is not a new topic. Already from early 90s it was clear that the benefits of replacing fossil fuel-based feedstock and reducing GHG emissions would lead to the need of additional land use and

related environmental impacts [14]. Therefore the LCA approach appeared immediately as a necessary tool for quantify all environmental impacts of biobased
materials in comparison with their conventional fossil fuel-based or mineral-based
counterparts in order to make strategic political and industrial decisions.

One of the earliest study is considered to be [15], which was commissioned by
The Swiss Federal Agency for the Environment, Forests and Landscape (BUWAL)
in order to have the first detailed and publicly available LCA for bio-based polymers [16]. Already at the beginning of the twenty first century, the review conducted by [16] counted twenty case studies focusing of LCA applications on
bio-based polymers or natural fibres. That review revealed a number of assumptions
about system models applied and data considered, which make weak results of the
assessments. Among others important findings, Ref. [16] pointed out that various
waste management treatment options should be included in LCAs for biopolymers
and natural fibres due to their strong impact on the final results.

Two other reviews of LCA studies on biomaterials that should be considered,
although they are not focusing on materials for agriculture, are [17] and [18]. In the
first paper, authors review readily available LCA studies or environmental
assessment studies for polysaccharide-based textile products, natural fibre composites and thermoplastic starch. As a result of the review, authors found that, for
each stage of the life cycle, polysaccharide-based end products show better environmental profiles than their conventional counterparts in terms of non-renewable
energy use and GHG emissions [17]. The second study [18] is focused on the
literature about the LCA of shoppers made from Mater-Bi$^{®}$ an polyethylene
highlighting, in general, a better performance of bioplastics in terms of reducing
consumption of non-renewable sources and emissions of greenhouse gases.

Later on, [14] conducted a review on the environmental impacts of bio-based
materials in a meta-analysis of 44 life cycle assessment (LCA) studies. In their
review emerged clearly that the variability in the results of life cycle assessment
studies made impossible drawing general conclusions. Ref. [14] highlighted that the
focus of early LCA studies was on initial focus on non-renewable energy use and
GHG emissions only, but other important impact categories (such as eutrophication
and acidification) should be included for a more realistic impact assessment. In fact,
as reported also in following studies [19] biobased materials may have higher
environmental impacts than their conventional counterparts in the categories of
eutrophication, acidification and stratospheric ozone depletion, as a result of the
agricultural practice for producing biomass.

The main issue related to the application of the LCA approach to biomaterials is
related to the fact that international guidelines for LCA usually do not specify the
calculation settings that should be specifically considered for case studies with
biomaterial [20] and therefore harmonized methods should be welcomed [21].

Most of the LCA studies focus on the production of bioplastics in the form of
pellets or resins, quite few are the case studies that include a LCA of products made
with bioplastics in their full life cycle, therefore including impacts from use phase
and disposal of the product. Among such studies very rare are the applications

related to agricultural products. To the knowledge of the authors just four papers have been published, in particular: (I) [22] calculated the Carbon Footprint to produce a *Petunia x hybrida* plant, from nursery phases to distribution, utilizing different pots such as petroleum-based plastic, bioplastic, rice hull, compressed peat and others; (II) [23] applied LCA to a nursery system comparing the real situation with a scenario in which 50% of all pots are from bio plastic and biodegradable; (III) [24] calculated energy consumption and GHG emission for strawberry crops grown in unheated plastic tunnels using currently existing cultivation techniques, post-harvest management practices and consumption patterns and the same strawberry cultivation chain in which some of the materials used were replaced with bio-based materials; (IV) [25] evaluated the environmental performance of two different mulch film systems (traditional and biodegradable mulch film). The last three papers will be discussed in the next sections of this chapter and, together with the results highlighted from review papers, a series of recommendations for LCA applications on biodegradable products and food for thoughts will be provided.

7.4   Key LCA Applications on Biodegradable Plastics for Agriculture

The worldwide plastic consumption in the agricultural sector was 1,150,000 t in 1985, 2,850,000 t (+147%) in 1999 and about 3,900,000 t in 2007 (+240% compared with 1985) of which 41% was related to mulch films [11]. In this section two case studies related to the most important applications of biodegradable bioplastic in the agriculture sector (i.e. mulch film and plots) are addressed, highlighting peculiarities in the environmental profiles assessments.

7.4.1   Case 1: LCA of Mulching Films

7.4.1.1   On the Agricultural Need on Mulching Films and Environmental Concerns

In all agricultural crops where such a practice is doable (from a techno-economic point of view), mulching provides significant agronomic advantages:

- increased yield (from 20% up to 60%) and higher quality of crops;
- weed control and reduced use of pesticides;
- early crop production (important for crops such as muskmelons, melons, watermelons) caused by the higher soil temperature;
- reduced consumption of irrigation water (up to 30% less water than for bare soil).

Nevertheless plastic mulch films need to be properly removed and properly disposed of at the end of the crop cycle. This implies collecting and recycling or, where this is not possible, landfilling or incinerating with energy recovery. The recovered mulch film is generally heavily contaminated with soil, stones and biological waste; this makes mechanical recycling difficult. In general, the contamination of mulch films represent from 50 to 75% of their initial weight [26]. In Italy, for example, it was estimated that, on average, about 400 kg of waste is produced per each mulched hectare, whose composition is 290 kg of mulch film and the rest soil, stones, vegetal residuals [11].

Occasionally (or in some areas in quite often) plastic films are not properly collected and recycled after their use but disposed of by burning in the field or by uncontrolled landfilling. Both practices cause environmental concern [27].

In some areas of the world a growing concern over the so called "*white pollution*" is becoming more and more evident. *White pollution* mainly refers to the negative effects on the environment of both "visual pollution" and "potential hazards", due to ineffective management practices of plastic waste. The "potential hazards" in agriculture are linked to the plastic film waste residues remaining in the soil, and which can, due to their accumulation year after year, lead to changes in the agriculture soil's characteristics as well as reduction in crop yields. Plastic pollution not only interest agricultural sector thus terra firma but also the oceans of the entire world, as shown by the latest researches addressing this phenomenon [28].

Many studies have been performed in China (i.e. Xinjiang Autonomous Region) on the effect of residuals of plastic mulch films in the soil, evaluating their overall impact (agronomical and on soil fertility). The soil structure, ventilation appeared to be damaged from the plastic residues, a general reduction in the soil moisture retention and movement of water in the soil were reported, resulting in a general decrease of the soil quality [29, 30].

The presence of plastic residues in the soil impact the growth and development of the crops (e.g. cotton), causing crop yield reductions up to 15%. The phenomenon as a whole is really relevant since in China the covered crop area is expected to reach above 30 millions of hectare, and residual plastic film mulch levels in the contaminated areas are in the order of 200 kg/ha in the top soil (0–20 cm) [11].

Since Asia represents the largest segment of mulch film demand in the global market where China accounts for about 70% of the regional demand [31], we can conclude that currently the "white pollution" is the most relevant environmental and social concern linked to mulch films use.

Its resolution needs joined interventions from policy, regulations and technology fields. The replacement of the current not biodegradable plastic mulch films with the biodegradables ones represent a valuable solution of the problem [11], especially in those rural areas uncovered by collection and recycling systems. Thanks to their biodegradation and absence of accumulation and toxicity effects [32], biodegradable mulch films do not need to be removed from the soil and disposed of, simplifying the farmers' operations. Starch based mulch films have been specifically designed for 1–9-month crops, and their processability is close to that of traditional plastics [33]. From a functionality point of view (i.e. during the cultivation phase), they show the similar positive effects of traditional plastic mulch films [34].

Based on literature review performed, the main constrains related to the large distribution of biodegradable mulch film does not rely on their functionality rather on the higher price compared to traditional plastic films. However, some studies conducted in Italy demonstrate that the overall economic expenditures for farmers, taking into account also the expenditures for mulch plastic film disposal, are comparable [35].

In the next section we will address LCA studies found in literature addressing plastic films. As pointed out in Sect. 7.2, LCA is currently not yet suitable to quantify environmental damage caused by white pollution nevertheless there are signals by the scientific community about the importance of a quantitative inclusion of plastic pollution (e.g. fragmentation in small pieces) within LCA methodology [36].

7.4.1.2 Environmental Outreaches of LCA on Mulching Films

The research described in [25] aimed to evaluate the environmental performance of two different mulch film systems (traditional and biodegradable mulch films) within the Italian context. The functional unit was set equal to "*1 ha of mulched agricultural land*" since, as reported above, there are no differences in the functionality (e.g. crop yield, quality of fruits etc.) of both mulch film systems. The main characteristics of the analysed products are shown in Table 7.1.

The characteristics of mulch films were derived from market info and data reflecting the Italian context. To fulfil the FU about 6000 m^2 of mulch film were needed, namely 96 kg of biodegradable mulch film and 288 kg of polyethylene mulch film per ha of mulched soil.

The end of life scenarios were set as follows:

- Biodegradable mulch film: 100% biodegradation in soil (a complete mineralization was assumed)
- Traditional mulch film: disposed according to average Italian (IT) disposal scenario for plastic waste coming from agricultural sector (i.e. 10% recycling, 76% landfill and 14% incineration).

Environmental credits for plastic mulch film were accounted for. These regarded: replacement of virgin polyethylene from recycling and heat and electricity production from incineration.

Table 7.1 Main characteristics of the analyzed mulch film

Mulch films characteristics	Biodegradable mulch film	Polyethylene (PE) film
Thickness (μm)	13	52
Weight (grams*m^{-2})	16	48
Colour	Black	Black

A sensitivity analysis regarding end of life scenario for plastic mulch film was performed as well (i.e. traditional mulch film disposed of through 100%, incineration, 100% landfill and 100% recycling). The "Cradle to grave" LCIA results for all analysed scenarios are shown in Table 7.2 (absolute values) and Fig. 7.1 (relative values).

The main outcomes of the study were:

- The granule production phase (not showed) of both materials dominates the environmental impacts especially for NRER and GW.
- The end of life treatments are significant for PE plastic mulch films (not showed).
- For impact categories like EU and AC polyethylene granule production has a lower potential impact compared to starch-based bioplastic material, however, since a higher amount of plastic is required for producing the PE mulch film and considering the compulsory removal and the disposal phases, the environmental performance obtained is worse whichever end of life scenario is considered.
- The worse scenario is represented by landfilling of non biodegradable mulch film since neither energy nor material is recovered
- The overall reduction of the potential impacts when biodegradable mulch film is used as an alternative to non-biodegradable plastic ranged from 25 to 80%.

Very rare are the LCA studies which focus on an actual field application of biobased plastic mulch films. The only paper available in the scientific literature is

Table 7.2 Life cycle impact assessment "cradle to grave" of 6000 m^2 mulch film (1 ha of mulched soil)

Impact category (*method used*)	Unit	Non-biodegradable mulch film				Biodegradable mulch film
		100% incineration	100% landfill	100% recycling	IT average scenario	100% biodegradation in soil
AC (*International EPD system*)	kg SO$_2$eq	1.83	3.22	**4.66**	3.16	1.40
EU (*International EPD system*)	kg PO$_4$eq	0.45	**1.91**	0.76	1.58	0.34
POF (*International EPD system*)	kg C$_2$H$_4$ eq	1.52	**1.66**	0.86	1.56	0.31
GW (*IPCC 2007 —100 years*)	kg CO$_2$eq	**1340**	849	1076	943	402
NRER (*IMPACT 2002+*)	MJ eq	20476	**25831**	19597	24436	5496
AD (*CML 2001*)	kg Sbeq	8.65	**11.08**	8.72	10.49	2.7

The figures in bold indicate the highest impact for each impact category. *AC: acidification potential, EU: eutrophication potential, POF: photochemical ozone formation potential, GW: global warming potential, NRER: Non-renewable energy resources, AD: abiotic depletion*

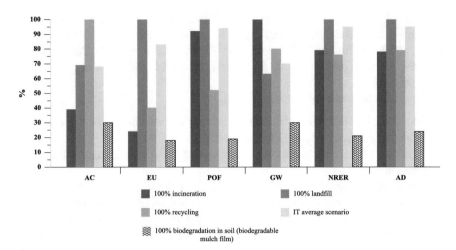

Fig. 7.1 Normalized LCIA results of Table 7.2 to the highest value (=100%)

by [24] which analyses strawberry industry in Northern Italy. The analysis was conducted using two scenarios as reference systems: strawberry crops grown in unheated plastic tunnels using currently existing cultivation techniques, post-harvest management practices and consumption patterns (base scenario) and the same strawberry cultivation chain in which some of the materials used were replaced with bio-based materials (alternative scenario), in particular mulch film (in the field stage) and food packaging (in the post harvesting phase). The scope of the systems examined encompassed the production of strawberries for fresh consumption, including post harvesting phases (i.e. refrigeration, packaging, distribution and packaging disposal). The functional unit for the purpose of reference was the consumer-unit, a 250-g flow pack.

The study assumed that for the base scenario 20% of plastic waste was incinerated and 80% landfilled, whereas for the alternative scenario it was assumed that 14% of plastic waste was incinerated and 86% composted.

The overall LCIA results for global warming potential and non-renewable energy resources for the analysed strawberry systems were 10–15% lower for the alternative scenario using starch-based bioplastics mulch film, compared to the base scenario using polyethylene one.

Compared to the analysis prepared by [25] the starch polymer's advantage compared to polyethylene is smaller (10–15% instead of 60–80%). This is mainly due to the fact that the analysis encompassed the whole supply chain of strawberry like their production, packaging, distribution and end of life. These phases contribute in the same way on both systems making the advantages of biodegradable mulch film replacement less significant. In any case 10–15% of reduction is still a valuable result considering that strawberry food supply chain require more energy for storage and packaging than other fruits and vegetables [37].

7.4.2 Case 2: LCA of Biodegradable Pots

7.4.2.1 On Plant Production Systems and Their Need of Plastic Materials

The production of flowers and potted plants is a very important agricultural sector: in 2012 it accounted an economic value of 26.5 billion Euro and Europe is one of the leading continents in the market, producing for 42.6% of the total economic value [38]. The first phase of the production of flowers and potted plants is nursery activities, conventionally divided into two main parts: the propagation area and the cultivation area.

The first is dedicated to the multiplication of plants, with the objective of providing guaranteed and healthy commercial material. This area is not present in all nurseries: some of them purchase raw propagation materials from other nurseries. The cultivation area is the largest nursery surface; it is divided into different sections, depending on the different species and on the development stage of the plants [39]. In both propagation and cultivation areas a lot of plastic is often used for several scopes (Fig. 7.2): to grow plants in containers, to isolate containers from soil using different types of mulching such as plastic semi-permeable sheets, to protect cultivation in tunnels or greenhouses.

About the use of pots, petroleum-based containers are largely the most used container type since the 1980s in both the United Sates [40] and Europe [23] and they are not use just for growing and transporting plants but also for marketing purposes, with the use of curious shapes and colours. Reference [40] noted that the relative low economic price of plastic contributed to establish the prominence of plastic containers in plant production systems, giving to containers a pivotal role for agricultural techniques and marketing strategies adopted by the nursery. Nowadays it is clear that the outcome of the combination of these factor is an overabundance of plastic waste which is not always possible to recycle for both technical reasons (the type of plastic used) and logistic issue (distance or accessibility to recycling centres).

The impressive use of plastic in nursery systems, heavily weight on the Climate Change potential of plant produced [39]. More precisely, Ref. [41] calculated the life cycle GHG inventory of an average tree produced in Monrovia Nursery (California) with the result of around 4.6 kg CO_2eq. per tree, the 16% of which (0.64 kg CO_2eq.) is due to plastic containers. This value was obtained already taking into account that the nursery was already reusing containers, therefore options for emission reductions related to a further increase of internal recycling were very limited.

Fig. 7.2 Two types of cultivation areas in nurseries: open air production in containers (**a**) (Vivaio Carlo Lazzerini, Cavaglià, BI, Italy) and protected cultivation (**b**) (Vivaio Purpurea S.N.C, Piobesi, TO, Italy). In both cases the soil is covered by plastic and plants are grow in pots

7.4.2.2 Environmental Outreaches of Biodegradable Pots Quantified by LCA

In several papers (e.g. [41, 42]), it is stated, without giving a specific extent, that great potential for environmental impact reductions in the case study could be gained from using containers manufactured with alternative lower-emission resin materials or by reducing polypropylene mass in containers. Nevertheless in order to properly assess if this estimation is exact a balance of benefits and drawbacks, in the light of LCA, has to be conducted.

There are several interesting researches about the agronomic performance of biodegradable pots for plant growing, but very rare are LCA studies (discussed later on). About the investigation of the agronomic performance, one of the key publications is [40] in which the current state of the art on the topic is presented. In the review it is pointed out that the use of biodegradable pots lead to lower the environmental burden because of their less impacting material, but also they involve a number of growing transitions in the cultivation technique because of their physical and chemical properties. In fact both positive and negative impacts of using biocontainers compared with petroleum-based containers have been reported on plant growth and development. In particular, most of the studies reported a more extensive root growth in plants produced in conventional plastic containers, compared with plant produced in traditional containers, with different extend from species to species [40]. This effect may be due to potential for water loss through biocontainer sidewalls because of the semi-porous nature of some alternative materials.

With the substitution of petrol-based pots with biodegradable plots, some changes in the agricultural practices are foreseen. As a consequence a full LCA should be addressed to quantify the environmental outreaches of such changes. At the knowledge of the authors, in just two papers [22, 23] a LCA approach is applied to nursery or horticultural production systems where biodegradable pots are used.

The study [22] applied a LCA to different production scenarios of *Petunia x hybrida* plant in order to quantify differences in energy consumption and GHG emissions for different containers. In particular the study applied a cradle-to-market approach in the quantification of the climate change potential of a single finished plant in a 10-cm diameter container. Emission form production and disposal of 10 container types were considered, among which: bioplastic, sleeve, rice hull, straw, manure and petroleum-based plastic as control. Authors of [22] quantified secondary impacts related to the increased need of irrigation and fertilisers due to the agricultural properties of the different pot materials, but—unfortunately—they did not considered GHG emission in producing alternative pots, stating that the work was just a preliminary study. Data for the compilation of the life cycle inventory were both primary (including interviews with producers and original findings from a series of experiments) and secondary. Results of the assessment quantified the petroleum-based plastic container as responsible for 0.544 kg CO_2eq, which correspond to 16% of overall climate change potential of the production cycle considered. Unfortunately authors are not able to quantify effective GHG emission variation that occur when using alternative materials for pots, but they highlight that indirect variations (due to the change of agronomic need because of different materials) are negligible. As a theoretical reflection we can highlight that if biomaterial used for pot manufacturing are by-products from agricultural systems just the impacts of the manufacturing processes should be considered and not the one of biomass production. This lead to a theoretical reduction of GHG emissions in comparison to petroleum-based plastic.

The study of [23] also accounted for climate change impacts alone in a nursery system, but also alternative impacts from the substitution of bio-based plastics for pots was considered. In detail the study apply an LCA, with a cradle-to-gate approach, to nurseries characterized by different production systems in the Pistoia district (Italy). The functional unit to which climate impacts are referred is 1 ha of nursery. This choice was taken because the nurseries considered in the assessment produce different kinds of plants [23], nevertheless using such functional unit alone, no information is given about the productivity of such systems, therefore a real comparison on the production performance in not possible. Results of the study highlight that the nurseries that apply cultivations in containers have higher climate impacts than the ones that adopt open field cultivations, with emissions ranged between 26.1 and 34.7 $MgCO_2$eq per hectare for the former, and between 2.3 and 6.6 $MgCO_2$eq per ha for the latter. The reasons for lower impact per ha of open field crop are mainly due to the avoided use of peat, plastic and electricity. In particular the relative weight of petroleum-based plastics on the climate impacts ranged between 15.9 and 34.6% of the whole nursery system. Authors of [23] calculated also the potential GHG emissions reduction according to the substitution of 50% of the traditional pots with biobased-plastic pot. The emission factor considered was 0.49 $kgCO_2$eq per kg of biomaterial (obtained from [43]) compared a range between 1.60 and 1.70 $kgCO_2$eq per kg of petroleum-based plastic (obtained from [44]). The result is an average net decrease of 5.2% of total GHG emissions of the systems using biobased-plastic plots. This study is therefore important for a preliminary quantification of the extent of the shift to bio-based pots in nursery. Unfortunately just variation in the climate change potentials of the scenarios have been assessed and no information is given in other impact categories.

7.5   Conclusions

Despite the great development of LCA methodology over the last 20 years, LCA studies on bio-based and biodegradable products used in agriculture sector are still limited. Nevertheless, it is possible to draw some important aspects and issues related to the environmental sustainability of biodegradable plastics in agriculture considering the general LCA literature and the two case studies previously described.

Firstly we have seen how LCA methodology, due to the high number of methodological choices and technical assumptions that characterized its execution, can make difficult attempts to generalize and compare LCIA results, suggesting the execution of LCA case by case. Secondly we have seen how LCA methodology is not yet suitable to quantify the potential impacts associated to "white pollution", the main concern of mulch film use, even if there are some indications among scientific community to do it. In particular, two of the main reasons for the unsuitability of the LCA approach (as so far) for white pollution are the lack of a proper impact category able to address properly such impacts and the lack of terms of reference (thresholds and normalization parameters) able to give a dimension of such impacts.

Coming to biodegradable mulch film LCA analysis, it was interesting to observe how it gained a better environmental performance also for those impact categories like acidification and eutrophication, where generally bioplastics score worse than traditional polymers (e.g. polyethylene). The reasons of that are: lower consumption of plastic required (thanks to the lower thickness), no removal and disposal phase as required for traditional mulch films. Overall the life cycle potential impacts reductions of biodegradable mulch film (alone) are in the order of 25–80% compared to traditional mulch film, depending impact categories considered. For example, for GW and NRER biodegradable mulch film resulted 60 and 80% respectively lower. In case of the whole supply chain (e.g. strawberry), the benefits of biodegradable mulch film become more contained (i.e. 10–15%) due to the fact the majority of the LCA strawberry phases (e.g. use of fertilizers, strawberry harvesting, distribution etc.) contribute in the same way on both systems flattening out the differences observed in the LCA of mulch film alone. Apart from LCA considerations, biodegradability of bio-based mulch film represents, for the authors, the heart of the matter of this application make these innovative products very effective for tackling the diffusion of plastics in the environment (i.e. white pollution), which represent the real life concern (e.g. Xinjiang Autonomous Region, China).

Biodegradable pots represent another interesting application where biodegradability should be better exploited so as to increase diversion from landfill. Unfortunately the rare cases available in technical and scientific literature consider only partially the added value of this product, as some life cycle benefits such as the direct transplanting of the pot (together with the plant) might also result in a direct saving of waste material, time and—in particular condition—also organic matter amending.

The LCA methodology, despite its flaws, is still the most systematic way to understand the interrelation between a product and the environment, nevertheless some specific issues, related to the own nature of biodegradable products, require a more detailed way to be properly addressed. We agree with [21], when they state that work is underway in many areas to develop more comprehensive and robust assessment tools to evaluate the environmental impacts of bio-based materials.

On top of all this, there is the assessment of challenges and opportunities of biorefineries. The latter will represent a milestone for bio-based products diffusion since, in the chemical industry, biomass represent the only non-fossil resource of carbon. There are robust demonstrations of the advantages presented by biorefineries for climate change protection, value creation and resource efficiency [45], as long as a complete and rationale utilization of biomass is achieved. In other words considerably technological advancement and innovation is necessary for the biorefinery concepts to be operated commercially and on industrial scale.

Concluding biorefinery interactions will increase complexity of the supply chains of bio-based products making the "big picture" not easy to define. Still some more years are needed to reach such methodological achievement.

References

1. Heijungs R, Suh S (2002) The computational structure of life cycle assessment, vol 11. Springer Science & Business Media
2. EU-European Commission (2010) International reference life cycle data system (ILCD) handbook e general guide for life cycle assessment e detailed guidance. Institute for Environment and Sustainability
3. Baumann H, Tillman AM (2004) The Hitch Hiker's guide to LCA. An orientation in life cycle assessment methodology and application. Studentlitteratur Lund, Sweden
4. Hauschild MZ, Huijbregts MA (2015) Life cycle impact assessment LCA compendium—the complete world of life cycle assessment. Springer, Dordrecht, Netherlands
5. Audsley E, Alber S, Clift R, Cowell S, Crettaz P, Gaillard G, Hausheer J, Jolliet O, Kleijn R, Mortensen B, Pearce D, Roger E, Teulon H, Weidema B, van Zeijts H (1997) Harmonisation of environmental life cycle assessment for agriculture. Final report concerted action AIR3-CT94-2028 European Commission DG VI Brussels, Belgium
6. Weidema BP, Thrane M, Christensen P, Schmidt J, Løkke S (2008) Carbon footprint. A catalyst for life cycle assessment? J Ind Ecol 12(1):3–6
7. Finkbeiner M (2009) Carbon footprinting—opportunities and threats. Int J Life Cycle Assess 14(2):91–94
8. Zamagni A (2012) Life cycle sustainability assessment. Int J Life Cycle Assess 17(4):373–376
9. Cerutti AK, Bruun S, Donno D, Beccaro GL, Bounous G (2013) Environmental sustainability of traditional foods: the case of ancient apple cultivars in Northern Italy assessed by multifunctional LCA. J Clean Prod 52:245–252

10. Hayashi K (2013) Practical recommendations for supporting agricultural decisions through life cycle assessment based on two alternative views of crop production: the example of organic conversion. Int J Life Cycle Assess 18(2):331–339
11. Liu EK, He WQ, Yan CR (2014) 'White revolution'to 'white pollution'—agricultural plastic film mulch in China. Environ Res Lett 9(9):091001
12. Notarnicola B, Salomone R, Petti L, Renzulli PA, Roma R, Cerutti AK (eds) (2015) Life Cycle assessment in the agri-food sector: case studies methodological issues and best practices. Springer
13. van der Werf HM, Garnett T, Corson MS, Hayashi K, Huisingh D, Cederberg C (2014) Towards eco-efficient agriculture and food systems: theory praxis and future challenges. J Clean Prod 73:1–9
14. Weiss M, Haufe J, Carus M, Brandão M, Bringezu S, Hermann B, Patel MK (2012) A review of the environmental impacts of biobased materials. J Ind Ecol 16(s1):S169–S181
15. Dinkel F, Pohl C, Ros M, Waldeck B (1996) Ökobilanz stärkehaltiger Kunststoffe (Nr 271) 2 volumes. Study prepared by CARBOTECH Basel for the Bundesamt für Umwelt und Landschaft (BUWAL) Bern Switzerland
16. Patel M, Bastioli C, Marini L, Würdinger E (2005) Life-cycle assessment of bio-based polymers and natural fiber composites. Biopolymers online
17. Shen L, Patel MK (2008) Life cycle assessment of polysaccharide materials: a review. J Polym Environ 16(2):154–167
18. Gironi F, Piemonte V (2011) Bioplastics and petroleum-based plastics: strengths and weaknesses. Energy Sources Part A: Recovery Utilization Environ Eff 33(21):1949–1959
19. Yates MR, Barlow CY (2013) Life cycle assessments of biodegradable commercial biopolymers—a critical review. Resour Conserv Recycl 78:54–66
20. La Rosa AD, Recca G, Summerscales J, Latteri A, Cozzo G, Cicala G (2014) Bio-based versus traditional polymer composites a life cycle assessment perspective. J Clean Prod 74:135–144
21. Pawelzik P, Carus M, Hotchkiss J, Narayan R, Selke S, Wellisch M, Weiss M, Wicke B, Patel MK (2013) Critical aspects in the life cycle assessment (LCA) of bio-based materials—reviewing methodologies and deriving recommendations. Resour Conserv Recycl 73: 211–228
22. Koeser AK, Lovell ST, Petri AC, Brumfield RG, Stewart JR (2014) Biocontainer use in a Petunia × hybrida greenhouse production system: a cradle-to-gate carbon footprint assessment of secondary impacts. HortScience 49(3):265–271
23. Lazzerini G, Lucchetti S, Nicese FP (2014) Analysis of greenhouse gas emissions from ornamental plant production: a nursery level approach. Urban For Urban Greening 13(3): 517–525
24. Girgenti V, Peano C, Baudino C, Tecco N (2014) From "farm to fork" strawberry system: current realities and potential innovative scenarios from life cycle assessment of non-renewable energy use and green house gas emissions. Sci Total Environ 473:48–53
25. Razza F, Farachi F, Tosin M, Degli Innocenti F, Guerrini S (2010) Assessing the environmental performance and eco-toxicity effects of biodegradable mulch films. In: VIIth international conference on life cycle assessment in the agri-food sector, Bari, pp 22–24)
26. Sorema (2008) Recycling schemes for thin mulching agricultural film analysis of the process and application examples. International Congress Plastic & Agricolture MACPLAS 2008. Bari, Italy 21–22 Feb 2008. http://wwwmacplas08org/
27. Garthe JW, Miller BG (2006) Burning high-grade clean fuel made from low-grade used agricultural plastics. Solid Waste Technol Manage 802–809
28. Eriksen M, Lebreton LC, Carson HS, Thiel M, Moore CJ, Borerro JC, Reisser J (2014) Plastic pollution in the world's oceans: more than 5 trillion plastic pieces weighing over 250000 tons afloat at sea. PLoS ONE 9(12):e111913
29. Dianjie N, Honge X, Liangsheng G, Dongmei Z, Haizhen Z, Pinghe R, Sbiwei C (1996) Study of the influence of the residue film on soil and cotton growth in the cotton fields. Acta Gossypii Sinica 1

30. Li F, Zhang L, Cui J, Dong H, Zhang C, Wang G (2005) Study of agricultural tri-dimension pollution on ecological system in cotton field and its control tactics. Cotton Sci 17(5):299–303
31. Reynolds A (2010) Updated views on the development opportunities in agricultural film markets. In: Proceeding of the international conference on horticultural and agricultural films and covers, Barcelona, 22–24 Nov 2010
32. Kapanen A, Schettini E, Vox G, Itävaara M (2008) Performance and environmental impact of biodegradable films in agriculture: a field study on protected cultivation. J Polym Environ 16 (2):109–122
33. Bastioli C, Facco S (2001) Mater-Bi starch based materials: present situation and future perspectives. In: Biodegradable plastics conference, Frankfurt, Germany, 26–27 Nov
34. Scarascia-Mugnozza G, Sica C, Russo G (2012) Plastic materials in European agriculture: actual use and perspectives. J Agric Eng 42(3):15–28
35. Cozzolino E, Leone V, Carella A, Piro F (2010) Mater-Bi® contro polietilene: più prodotto costi equivalenti. L'informatore Agrario 27
36. LCA discussion list personal communications Tuesday 31st of March 2015 and Wednesday 1st of April 2015. http://wwwpre-sustainabilitycom/lca-discussion-list
37. Cerutti AK, Beccaro GL, Bruun S, Bosco S, Donno D, Notarnicola B, Bounous G (2014) Life cycle assessment application in the fruit sector: state of the art and recommendations for environmental declarations of fruit products. J Clean Prod 73:125–135
38. European Commission DG Agriculture and Rural Development (2013) Live plants and products of floriculture. Working document of the Advisory Group Flowers and ornamental plants. http://eceuropaeu/agriculture/fruit-and-vegetables/product-reports. Accessed 220414
39. Russo G, De Lucia Zeller B (2008) Environmental evaluation by means of LCA regarding the ornamental nursery production in rose and sowbread greenhouse cultivation. Acta Hort 801 (2):1597–1604
40. Nambuthiri S, Fulcher A, Koeser AK, Geneve R, Niu G (2015) Moving toward sustainability with alternative containers for greenhouse and nursery crop production: a review and research update. Hort Technol 25(1):8–16
41. Kendall A, McPherson EG (2012) A life cycle greenhouse gas inventory of a tree production system. Int J Life Cycle Assess 17(4):444–452
42. Beccaro GL, Cerutti AK, Vandecasteele I, Bonvegna L, Donno D, Bounous G (2014) Assessing environmental impacts of nursery production: methodological issues and results from a case study in Italy. J Clean Prod 80:159–169
43. Harding KG, Dennis JS, Von Blottnitz H, Harrison STL (2007) Environmental analysis of plastic production processes: comparing petroleum-based polypropylene and polyethylene with biologically-based poly-β-hydroxybutyric acid using life cycle analysis. J Biotechnol 130(1):57–66
44. Plastic Europe (2005) Eco-profiles of the European plastic industry. http://wwwplasticseuropeorg/plastics-sustainability-14017/eco-profilesaspx
45. Federal Government of Germany (2012) Biorefineries roadmap (as part of the German Federal Government action plans for the material and energetic utilisation of renewable raw materials). (www.bmbfde/pub/roadmap_biorefineriespdf)

Printed in the United States
By Bookmasters